International Scientific-Technical Conference on Innovative Engineering Technologies, Equipment and Materials 2015 (ISTC-IETEM 2015)

IOP Conference Series: Materials Science and Engineering
Volume 134

Kazan, Russia
2 - 3 December 2015

ISBN: 978-1-5108-2673-1
ISSN: 1757-8981

Printed from e-media with permission by:

Curran Associates, Inc.
57 Morehouse Lane
Red Hook, NY 12571

Some format issues inherent in the e-media version may also appear in this print version.

Printed by Curran Associates, Inc. (2016)

For permission requests, please contact the Institute of Physics
at the address below.

Institute of Physics
Dirac House, Temple Back
Bristol BS1 6BE UK

Phone: 44 1 17 929 7481
Fax: 44 1 17 920 0979

techtracking@iop.org

Additional copies of this publication are available from:

Curran Associates, Inc.
57 Morehouse Lane
Red Hook, NY 12571 USA
Phone: 845-758-0400
Fax: 845-758-2633
Email: curran@proceedings.com
Web: www.proceedings.com

Table of contents

Volume 134

International Scientific-technical Conference on Innovative Engineering Technologies, Equipment and Materials 2015 (ISTC-IETEM-2015)
2–3 December 2015, Kazan, Russia

Accepted papers received: 3 June 2016
Published online: 8 July 2016

Preface

Papers

IOP Conf. Series: Materials Science and Engineering **134** (2016) 011001 doi:10.1088/1757-899X/134/1/011001

The International Scientific Conference "Innovative engineering technologies, equipment and materials – 2015" took place at the exhibition center "Kazan Fair" within the framework of the 15th international specialized exhibition "Mechanical engineering & Metalworking" and the 10th specialized exhibition "TechnoWelding" on the 2–3 December, 2015.

The conference program included 5 sections:

Section 1 - "Innovative technologies in the design and manufacture of engineering products. Practice and prospects of creation and application"

Section 2 - "Innovative development of mechanical engineering and economics"

Section 3 - "Mathematical and physical modeling of information, technical, technological and managerial systems and processes"

Section 4 - "Innovative development of small and medium-sized enterprises"

Section 5 - "Innovative welding technology in the industry"

Participants and guests of the conference discussed the scientific and technical achievements in the field of development, design and manufacture of engineering products.

The main results of studies of the country's leading universities in the field of robotics, new materials (including the production of waste), mechanical engineering, energy and economy were presented.

The Presidium included Director of the Institute of Engineering Prof. Nail F. Kashapov, Director of Innovation of "KNIAT" Peter P. Baskevich, Director of educational activities of Engineering Institute, Idar I. Hafizov and Director of the Integrated Learning Center for Energy Efficiency, Iskander R. Gilmanshin.

Research and innovation at the Kazan Federal University was presented by the staff of the Engineering Institute, Naberezhnye Chelny Institute, Chemistry Institute (named after Butlerov) and the Institute of Management, Economics & Finance.

Speakers provided practical ways to improve performance for the sixth technological order by Boris M. Morozov (PJSC "KAMAZ"); on the example of JSC "90 pilot plant", R.T. Sirazetdinov (KNRTU KAI / 90 pilot plant) examined the problem of sustainable development of engineering enterprises, E.Y. Kurchatov presented a new type of energy-efficient flywheel energy storage, and Y. I. Azimov and I. R, Gilmanshin presented a resource-recycling technology for tire production.

For postgraduate students the experience of live communication with scientists, practitioners, representatives of the real economy will allow future experts to determine its scientific direction for a successful solution of important problems of import substitution and enhancing the competitiveness of the engineering industry of our country.

Peer review statement

All papers published in this volume of *IOP Conference Series: Materials Science and Engineering* have been peer reviewed through processes administered by the proceedings Editors. Reviews were conducted by expert referees to the professional and scientific standards expected of a proceedings journal published by IOP Publishing.

IOP Conf. Series: Materials Science and Engineering **134** (2016) 012001 doi:10.1088/1757-899X/134/1/012001

Development of water saving technology for water supply system of industrial enterprises

A A Akhmetov, R V Akhmetova

Kazan State Power Engineering University
51, str. Krasnoselskaya, 420066, Kazan, Russia

Annotation
The article describes a variant of a wastewater treatment enterprise with the subsequent return of treated water into the water system.

According to the state policy in the field of environmental development of the Russian Federation for the period up to 2030 the strategic goal is defined as the solution of social and economic problems, providing an environmentally-oriented economic growth, preservation of favorable environment. Implementation and use of low-waste and resource-saving technologies and equipment to solve the problem to ensure environmentally sound waste management is required.

One of the most important issues of environmental protection is to protect the drainage basin from contamination. Important measures for the protection of water sources include industrial wastewater purification and further use of wastewater for industrial water supply.

Water-use efficiency in industry is estimated by the percentage of water used in the reverse system P_{rev}, rational use factor of water taken from the source K_{ru}:

$$P_{rev} = \frac{Q_{rev}}{Q_s + Q_{rev} + Q_m} \cdot 100 \Rightarrow 100\%,$$

$$K_{ru} = \frac{Q_s + Q_m - Q_{сбр}}{Q_s + Q_m} \leq 1,$$

where Q_s - amount of water taken from the source, Q_{rev} - the amount of water used in the back line of the system, Q_m - the amount of water coming from the raw materials.

The most dangerous wastewater is produced in the chemical and petrochemical industry. Wastewater of such enterprises is characterized by a complex and variable composition, high content of dissolved impurities, hence biological methods are not always effective. To reduce the wastewater discharge into the sewage system and water bodies is proposed to apply the thermal treatment of wastewater with partial or full reuse of purified water in reverse cycle industrial water supply system. The purpose of the thermal treatment of wastewater is a termination of its discharge with producing a refined distillate at the same time.

IOP Conf. Series: Materials Science and Engineering **134** (2016) 012001 doi:10.1088/1757-899X/134/1/012001

It is proposed to consider the process which includes three stages of wastewater treatment with the subsequent return of treated water into the water system (mainly for system replenishment):

Stage 1- removal of mechanical impurities (use of hydrocyclones, mechanical filters, ponds, lagoons, etc.);

Stage 2 - the use of evaporation plant for the concentration of wastewater;

Step 3 - a dry residue of impurities in the spray dryer, followed by recycling and disposal.

Schematic diagram of a zero-discharge water system with heat treatment of wastewater is shown in fig. 1.

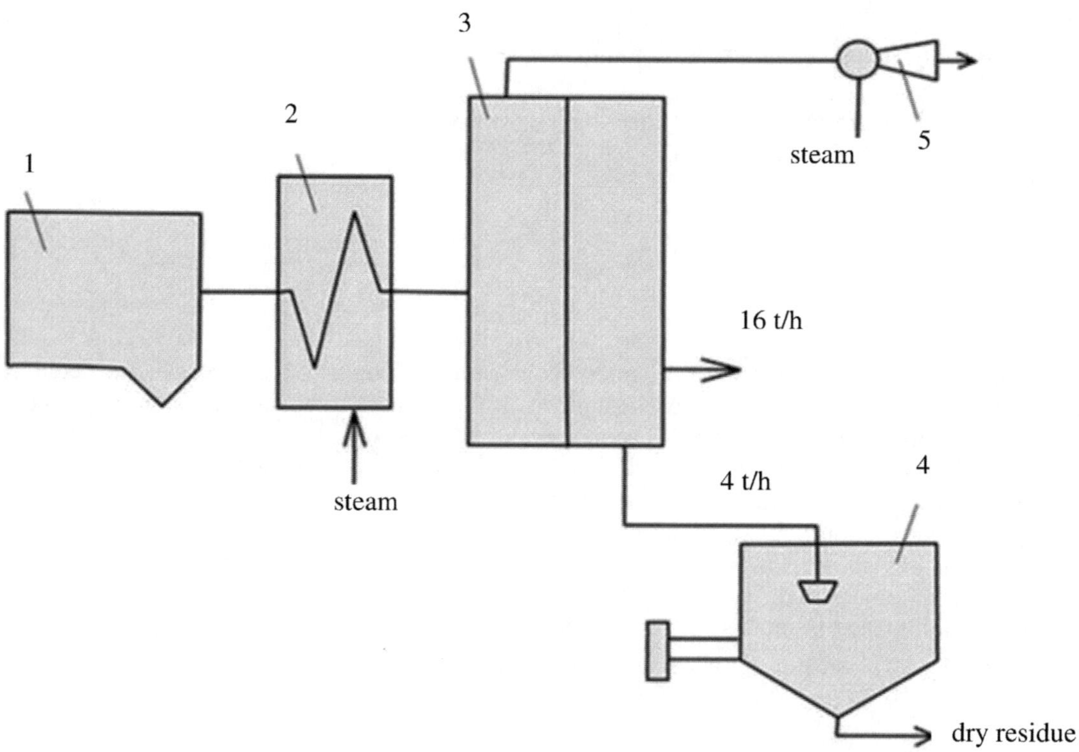

Fig. 1. A zero-discharge water system scheme:
1 - hydrocyclone; 2 - heat exchanger; 3 - instant-boiling evaporator; 4 - spray dryer; 5 – ejector

Wastewater goes into a hydrocyclone, where suspended sludge particles are separated from the aqueous suspension, after that the effluents are heated to a temperature of 90 ° C and fed to the instant-boiling evaporator (IBE). To improve thermal efficiency EIB has multistage configuration with subsequent change of water temperature from 90 to 40 ° C. Each stage of EIB includes following structural elements: the expander, the separator and the condenser (fig. 2).

The main advantages of using instant-boiling evaporators for obtaining of concentrate in the process of industrial wastewater treatment are:

- The absence of moving parts in the EIB, which leads to a safe operation;
- Highly qualified personnel are not required for the EIB maintenance;
- Installation can be operated in automatic mode;

- The quality of the distillate obtained corresponds to the quality of the condensate and it can be reused in the water supply system of industrial enterprise;
- Secure water purification from various types of pollution. Dissolved and suspended substances are removed from the water; mineral and synthetic emulsions are decomposed
- Dimensions of instant-boiling evaporator allow it to be incorporated into the wastewater treatment scheme without high capital costs.

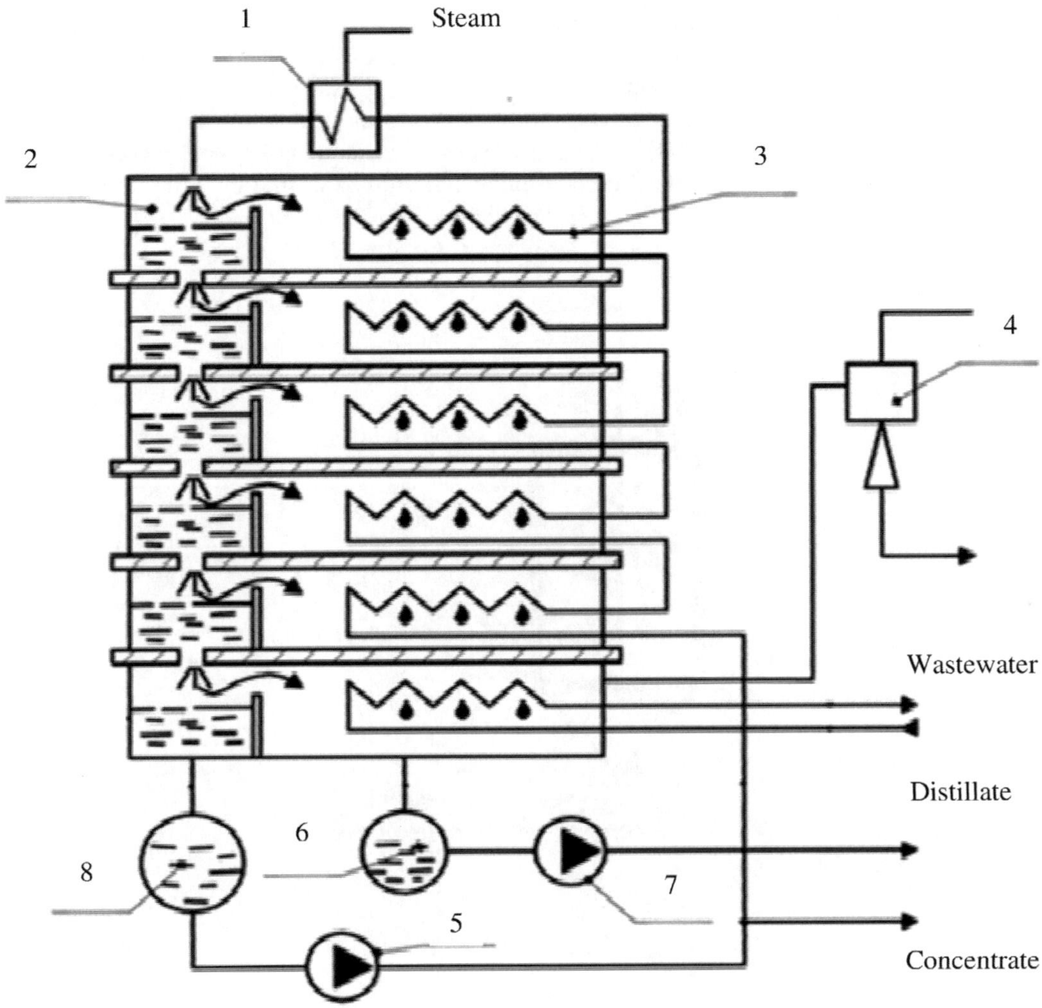

Fig. 2. Instant-boiling evaporator:
1 - heater; 2 - expansion chamber; 3 - condensation chamber; 4 - ejector; 5 - circulation pump; 6 – tank for distillate; 7 – distillate pump; 8 – tank for circulating water

Instant-boiling evaporator comprises a housing, a vertical partition wall with an opening for the passage of steam. The partition wall is tightly coupled to the housing and divides the evaporator into expansion chamber and condensation chamber. Evaporator also comprises pipes for supply and discharge of evaporating water and distillate. Condensation chambers and expansion chambers are equipped with independent bottoms by the number of cameras and tightly connected to the partition wall and the housing of the evaporator. The partition wall is additionally equipped with openings for the passage of steam by the number of expansion chambers and bottoms house cylindrical tubes for the passage of fluid. This allows the performance of the evaporator and reliability of its operation to be improved by eliminating the flow of highly mineralized water to the distillate and preventing contact between the corrosive environment and welded joints on the body.

Boiling does not occur on the heated surface, but in the free volume of the expansion chamber. This, combined with a multistage process of evaporation of the original water, can reduce the risk of scale accumulation in the internal volume of the device. Evaporator blowdown, depending on the quality of the wastewater, takes from 5 to 20% of plant capacity, and the yield of distillate is from 95 to 80% of entered volume. Blowdown water (concentrate) is supplied from the EIB to a spray dryer (fig. 3). The concentrate passes through a high-speed rotary atomizer in the top of the plant and atomized into fine droplets as a mist. The concentrate is dried instantaneously upon contact with the heated air and converted to powder. Solids are continuously removed from the bottom of the unit by a screw.

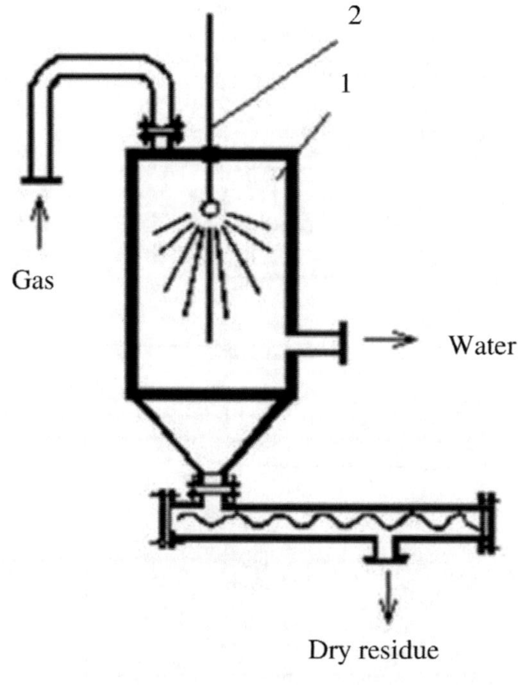

Fig. 3. Spray dryer:
1 - drying chamber; 2 - nozzle.

Nature of the drying process of materials in an atomized state is that the dispersed liquid droplets or fluid-like mass during its propagation in a confined volume is dewatered due to the difference in partial vapor pressure on the liquid droplets surface and in the environment. Depending on the technological requirements to the material either powder or plastic mass can be obtained in the spray dryer.

In comparison with other methods of drying of liquid and fluid-like materials spray drying has the following advantages:

IOP Conf. Series: Materials Science and Engineering **134** (2016) 012001 doi:10.1088/1757-899X/134/1/012001

- Formation of a significant interaction surface between the disperse phase and dispersion medium;
- Short-term process;
- Obtaining of a granulated powder material;
- Mechanization and automation of the drying process.

The spray drying process can be subdivided into three stages: spraying mass; heat and mass transfer between the droplets (particles) and the environment; the allocation of the dried product from the gas stream. Such division of the process is somewhat arbitrary, since it is impossible to identify a clear boundary between the stages due to their imposition of one another.

In the most general case, the spray process is meant to be the liquid jet fragmentation into a large number of drops and the allocation of these droplets in the volume. Fragmentation of the liquid jet into drops is a very complex process due to a number of external and internal factors. The main external factor is considered to be the impact of the aerodynamic force on the jet surface tending to deform and break it. Internal factors are various kinds of initial perturbations associated with the construction of the sprayer, the quality of its production, the turbulent motion of the liquid in the sprayer and so on. During atomizing the liquid jet is divided into a number of droplets with different diameter.

Methods of obtaining dry residue from the concentrated solution depend on the type of drying facility used.

Replacement of natural waters used for technological and energy needs by treated wastewater will solve the problem of the elimination of the deficit of water resources and prevent the depletion of stocks.

Selecting the method of wastewater treatment with subsequent return of the treated water into the water supply system is determined by the quality of water for the needs of production and the feasibility study of the adopted scheme.

References

[1] Fundamentals of the state policy in the field of environmental development of Russia for the period up to 2030 (Approved by. The President of the Russian Federation on April 30 2012)

[2] Pokrovskij V N and Arakchev E P 1980 *Wastewater thermal electron power plant* (Moscow: Energy) p 256

[3] RD 153-34.1-42.102-98 Guidance on designing thermal distillation units and evaporators for wastewater treatment of thermal power plants and thermal condensing power plants

[4] Konahina I A, Konahin A M, Ahmetov A A and Fazullina A I 2007 *Improving the efficiency of systems for the collection of condensate and water recycling in major petrochemical plants* (News of higher educational institutions. Problems of power engineering) 5-6

[5] Strizheus E V 2008 *The use of vacuum distillation equipment for the treatment of pro-industrial waste* (Ecology of production) 2

The analysis and selection of methods and facilities for cutting of naturally-deficit materials

I D Akhmetov, A R Zakirova and Z B Sadykov

Kazan (Volga region) Federal University
18, ul. Kremlevskaya, 420008, Kazan, Russia

Abstract

The comparison of perspective methods is done in the article, such as laser, plasma and combined electro-diamond methods of hard processed materials cutting. There are the review and analysis of naturally-deficit materials cutting facilities. A new electrode-tool for the combined cutting of naturally-deficit materials is suggested. This electrode-tool eliminates electrical contact between the cutting electrode-tool and side surfaces of the channel of cutting workpiece cut, which allows to obtain coplanar channels of cut.

Introduction

The rapid development of mechanical engineering, instrumentation, aviation, space industry, led to widespread use of hard, natural-deficit materials and their compositions, such as wolfram, cobalt, molybdenum, zirconium, niobium, as basic materials in the manufacture of load-bearing structures of products, where it is necessary to ensure high standards of accuracy and surface quality at an acceptable process performance.

Efficient processing of output parameters of technological processes of obtaining products from the hard and naturally-deficit materials is an important task in the conditions of modern production.

Analysis and comparison of separating materials methods

In modern industry for cutting natural-deficit, hard metals there are used traditional, such as mechanical processing, and new kinds of cutting, such as ultrasound, laser, plasma, and others [1].

Laser and plasma cutting of materials are the most modern and advanced and perspective methods which are being found more in practical use in the manufacture of hard materials.

These methods provide high efficiency of the production process and the possibility of cutting the curved contour of the workpiece, but have some disadvantages that prevent their use in the case of stringent requirements on the quality of treatment. For example, in laser cutting the grooves of certain depth, frequency and shape could occur, which together defines microgeometry of cutting area unallowable in the specific conditions of process requirements. In the case of treatment at low speeds there is the possibility of cutting mode automatically (autogenous) cutting, when there is a significant increase in the width of cut caused by the diameter of the jet of oxygen and not by the diameter of the focused laser beam.

During laser cutting it is also a probability of burr occurring on the lower edge of the workpiece as a result of surface wetting of the solid metal liquid metal and its oxide. Quantity of burrs is depended on parameters such as the power of the radiation, the thickness of the metal, the cutting speed and the pressure of the oxygen. Moreover, it is difficult to achieve minimal heat affected zone by laser cutting [2].

During laser cutting of large thickness workpieces there is slag can be formed in the cutting zone caused by the presence of alloying elements in the material being processed. The alloying elements in compositions influence the melting point of both the composition and its oxide. In some cases, refractory oxides of the alloying elements is above the melting temperature of the base material blank, which prevents the cutting process.

As well one of the technological features of laser cutting of large thickness materials is a need for deepening the focal spot of the radiation to cut the workpiece. The consequence of deepening of the focus is the increase of the hole diameters at the inlet to cauterize the material and therefore increase the gas inside the material in the melt zone [3].

A plasma cutting is another method of separation of materials which is used in modern production. The plasma cutting process consists in the local intensity of melting metal to be cut in the volume of the cavity cut by heat generated compressed arc, and removing molten metal from the cavity high speed plasma stream resulting from the channel of the plasma torch nozzle. During plasma cutting the quality and performance of the processing are depended on the composition of a plasma environment.

The disadvantages of plasma cutting are: low resistance of electrodes; possibility of saturation of the surface of cut by air gases; dependence of deterioration of consumables cleanliness and air humidity, the thickness of metal to be cut and value of operating current. The best indicators of plasma cutting is achieved by plants, which are equipped with a contour copy, tracking and control of technological parameters of the process of cutting devices to stabilize the working distance of the plasma torch. Such devices and systems are expensive, and in many cases their use is inappropriate for cutting difficult-to-deficient and naturally-deficit materials [4].

Based on an analysis of the advantages and disadvantages considered above methods it can be concluded:

- when cutting metal materials with thickness less than 6 mm it is mostly efficient to use laser cutting;
- when cutting metal materials with thickness more than 20mm it is more efficient to use plasma cutting;
- due to the difficulty to ensure the constancy (stability) of the parameters of laser and plasma cutting changes occur in the material layer at a considerable depth after treatment;
- possibility of a grid of cracks due to stress-strain state in the thermal overvoltage;
- use of laser and plasma cutting in the manufacture of products from naturally-deficit materials is undesirable because of the increased amount of material due to remove the altered layer;
- from economical point of view the most acceptable method is the use of combined electro-diamond cutting method, because the cutting is done with a thin disc, which is less than 1 mm and has diamond coating.

To illustrate that the proposed combined electro-diamond method is the most effective method for the separation of naturally-deficit and hard materials, there are shown in Table 1 the main differences in quality of cut between laser, plasma and combined electro-diamond cutting [5]:

Table 1. The main differences in quality of cut between laser, plasma and combined electro-diamond cutting methods.

Indicator of quality	Laser cut	Plasma cut	Combined electro-diamond cut
Taper edge	0–2	0–10	0
The surface roughness R_a, mcm	1,25–2,5	6,3–12,5	0,03–0,05

Reflow frames, corners	min	present	absent
Slag	min	absent	absent
Grid cracks	possibility	possibility	absent

Facilities for cutting of naturally-deficit materials

Technical problems solving of cutting deficit and hard materials by using combined electro-diamond cutting method are reflected in the derived with the participation of the authors of patents for inventions SU 1016129, SU 1641539, SU 1641540, RU 2432240, and used in the development of the machine SR 902M that enables adaptively manage the process of cutting in overall and individual components of the combination of mechanical, electrical discharge, electrochemical, hydraulic and cavitation treatments.

The cutting process of the machine starts with the formation of holes (like the punching before drilling) on the author's certificate SU 1016129. When applying the electrolyte in the electrode gap and the voltage of the additional current source available on the machine with adjustable law V = k * h, where V - voltage, h - electrode gap, k - empirical coefficient equal to 70V / mm [6].

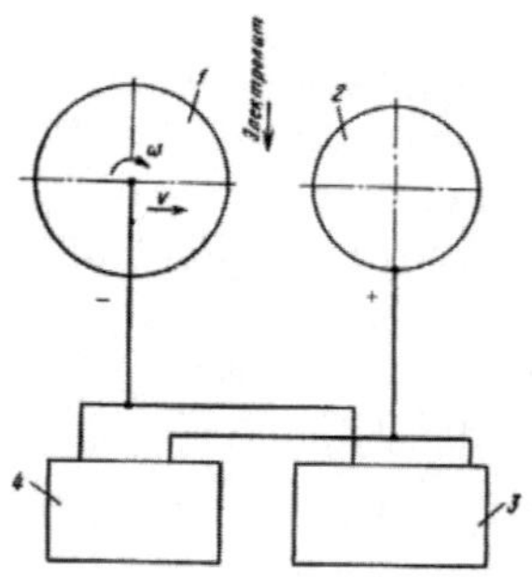

Figure 1 Installation scheme by inventor's certificate SU 1016129

Figure 1 shows an installation scheme for carrying out the proposed method. The scheme includes an abrasive tool 1, workpiece 2, source of operating voltage 3 power supply 4.

Stable introduction of a rotating tool made of thinsheet material, the body blank being cut, an apparatus is described in the inventor's certificate SU 1641539. The apparatus comprises a disc electrode-tool 1 for cutting the workpiece 2, the injector 3, inverters 4 beats, beats secondary converter 5, the generator 6 Supply pressure regulators 7, 8 technological current source (Figure 2).

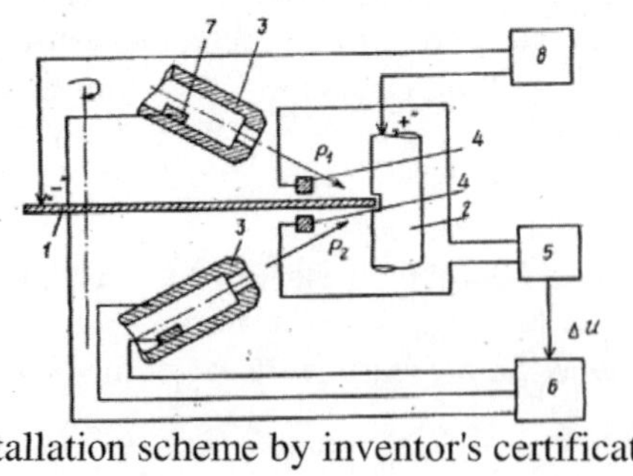

Figure 2. Installation scheme by inventor's certificate SU 1641539

The essence of the invention lies in the fact that to improve cutting precision due to the decrease of the beat tool electrode, an electrolyte pressure change by a signal proportional to the amplitude of the beat converter beats tool electrode through one nozzle - in phase and in antiphase via another.

The disadvantage of this invention is the fact that the processing of the material produced single cutting wheel, being in constant contact with the workpiece that does not provide an intensive electrochemical dissolution of the surface of the material and the evacuation of the dissolution and removal of beating tool electrode, then all this leads to rapid deterioration of electrode-tool and as a result in poor cutting (processing). Another drawback is the complex system of converters and regulators [7].

To improve the processing conditions by increasing the amount of electrolyte in the treatment zone and strengthening the electrochemical metal removal and cavitation components of the combined cutting process proposed "Extra electrode tool cutting electro-abrasive" copyright certificate SU 1641540, containing a set of disks of 1.2 combined with each other eccentrically relative to the axis of rotation of the tool-electrode (see Figure 3).

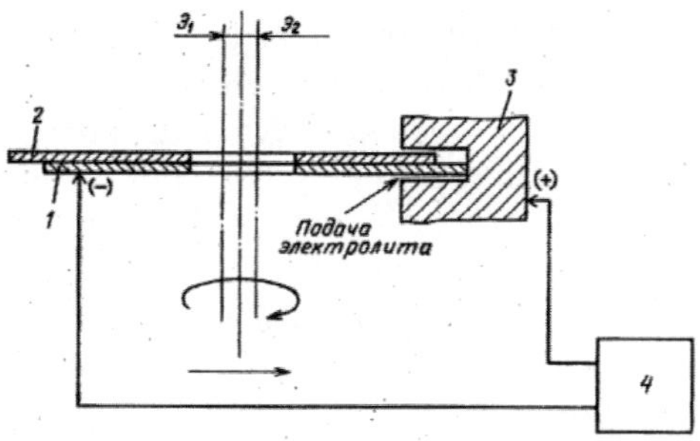

Figure 3. Installation scheme by inventor's certificate SU 1641540

Figure 3 shows a diagram of the device's certificate SU 1641540, where 1 and 2 - discs (an electrode-tool), 3 - blank, 4 - current source technology, A1 and A2 axis of the instrument.

The disadvantages of this device the collecting electrode tool is the following: when introducing an instrument into the body of the preform there is a "bump" plunge, which causes vibration of the tool and the workpiece, which reduces the quality of processing, and can also cause damage as the tool and the workpiece that generally leads to a decrease in the quality of processing [8].

To exclude the impact the introduction of the instrument into the body of the preform proposed collecting electrode-tool combination elektroalmaznoy cutting Copyright certificate RU 2432240. Extra electrode composite tool made in the form of a set of combined eccentric disc is characterized in that a set of eccentric disks introduced an additional drive, which is located relative to the axis of rotation of the tool-electrode without eccentricity and mounted between the disks with the possibility of permanent mechanical contact with the workpiece, allowing to produce smooth infeed, the eccentric wheels located on mutually perpendicular axes (Figure 4).

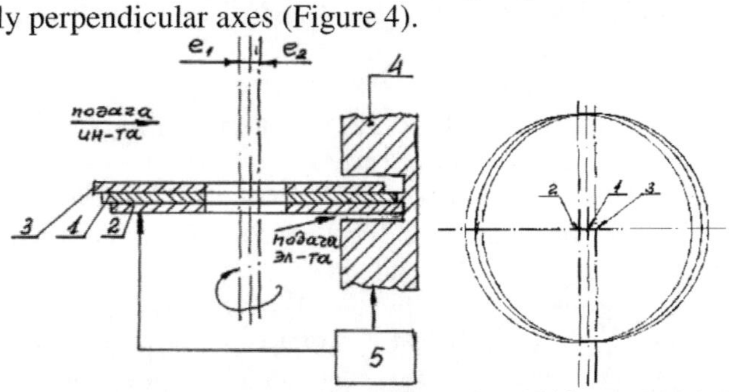

Figure 4. Installation scheme by inventor's certificate RU 2432240

Figure 4 shows a diagram of the device's certificate SU 2432240, where 1 - a disc without eccentricity, 2 and 3 - wheels with eccentricity, 4 - preparation, 5 - power supply E1, E2 - the axis of rotation of the tool.

Because of the disc 1 without eccentricity, located in the center of the disks with an eccentricity of 2 and 3 is provided by constant contact between the disk 1 and the workpiece 4, it promotes "unstressed" (smooth) plunging alternately drives 2 and 3 of the workpiece. When this vibration tool electrode does not occur.

Disadvantages tool electrode Copyright certificate RU 2432240 are: the introduction of an instrument into the body of the preform, an electric interaction between the side peripheral metal conductive strip holding diamond granules and cut the side surface of the workpiece, which causes non-uniform electric metal removal and receipt of the inclined cut surface, which reduces the surface quality and requires additional processing and increases the complexity of the process [9].

New electrode-tool for naturally-deficit materials cutting

The above illustrated methods have given some positive results, but does not completely eliminate the problem of uncontrolled removal of metal from the side surfaces of the channel cut. Design features of the buildings considered diamond discs (whole-metal material) does not allow the process to locate an electric pickup to a point and carry out the process at the minimum clearances.

We are offering fundamental solution to the problem of an uncontrolled removal of metal from the side surfaces of the channel of cut. The essence of the new method is described in the patent's application "Electrode-tool for combined cutting of conductive materials." The design of a new electrode-tool has nanomaterials and composites. In the present option, there is no uncontrolled permanent electrical contact between the electrode-tool and the workpiece, so that the side surface of the channel obtained by coplanar cut.

The principal difference between our method from the existing ones that due to the design of the proposed tool, it excludes side etching, which is present in all previous studies. Due to the presence in the electrode-tool conductive composites of carbon fiber processing is provided on a soft mode (microcurrents). A positive result of a new electrode-tool is the exclusion of additional processing operations to eliminate uncoplanar surface channel cut that does not require additional costs.

Bibliography

[1] Z B Sadykov, A F Mukhametzyanova, D F Graph 2015 Analysis and selection of an effective method of cutting and improvement electro-diamond method of hard cutting and natural-deficit materials Problems and prospects of development of aviation, ground transportation and energy "ANTE-2015": The International Scientific Conference, Kazan: Publishing house "Brig" pp 169-175

[2] Kostromin S V 2014 Features laser cutting low-carbon steel sheet Collection of scientific works of the 4th International Scientific and Practical Conference "Innovation, quality and service in technique and technologies", Kursk .: Publishing House ZAO "Universitetskaya kniga" pp 310-312

[3] V P Alekhin, A A Zverev, A M Lubchenco 2006 Laser cutting stainless steels large thicknesses Proceedings of the Moscow State Industrial University 4 (5) pp 13-15

[4] A N Vasiliev, V V Grandson, V I Zinoviev, T V Kotkina 2014 Comparative studies of air plasma and oxygen plasma cutting Proceedings of Moscow State Technical University MAMI 2 (20) pp 13-18

[5] Can plasma cutting compete with laser: website of "Gigameh". URL: http://www.gigamech.com/info-mmi/articles-mmi/92-lazer-vs-plasma (date of use 11/20/2015)

[6] Pat.1016129 USSR V23R1 IPC / 10. Electrical-abrasive processing method in an electrolyte medium [Text] / Smolentsev V.P., Sadykov Z.B., Gafiatullin S.S. .; Number 3318916 / 25-08; appl. 07/17/1981; publ. 05.07.1983, Bull. № 17. - 2.

[7] Pat. 1641539, the Soviet Union, the IPC V23H7 / 12 V23H7 IPC / 36. A method of cutting electric methods [Text] / Odintsov I.A., Tarasov S.F., Guzhavina V.B., Sadykov Z.B., Smolentsev V.P. .; № 3964776/08; appl. 17.10.1985; publ. 15.04.1991, Bull. Number 14 – 2 p.

[8] Pat. 1641540, the Soviet Union, the IPC V23H7 / 12 V23N5 / 10. Extra-electrode tool cutting elektroabrazivnoy [Text] / Odintsov I.A., Tarasov S.F., Sadykov Z.B., Zorohin A.F., Smolentsev V.P. .; №4117955 / 08; appl. 04/06/1986; publ. 15.04.1991, Bull. Number 14 (II h.). - 2 p.
[9] Pat. 2432240 Russian Federation V23H7 IPC / 12 V23N5 / 10. Extra-electrode tool cutting elektroalmaznoy [Text] / Sadykov N.Z., Sadykov Z.B., Zakirov A.R. .; applicant and patentee GOU VPO KSTU. AN Tupolev. - № 2010105564; appl. 16.02.2010; publ. 27.10.2010 – 3 p.

IOP Conf. Series: Materials Science and Engineering **134** (2016) 012003 doi:10.1088/1757-899X/134/1/012003

Modern technologies of waste utilization from industrial tire production

Yusuf Azimov[1], Iskander Gilmanshin[1], Suriya Gilmanshina[1]

[1] Kazan (Volga region) Federal University, 420008,
Russia, Kazan, Kremlevskaya street, 18

e-mail: is-er@yandex.ru

Abstract. The innovative technology of waste tire production recovery from JSC "Nizhnekamskshina", which determines the possibility of obtaining a new type of composite material in the form fiber filled rubber compound (FFRC) as the raw material, production of rubber products with high technical and operational characteristics.

The development of modern waste technology from industrial production determines to comply the federal laws such as "About environmental protection" (Federal Law №219 from 21.07.2014),"About energy saving and energy efficiency improvements and about amendments to certain legislative acts of Russian Federation" (Federal Law №261 from 23.11.2009).

In the present project technological wastes are considered as secondary material resources used in the recycle mode of technological processes of redistribution with production of high quality products.

The problem of waste production is relevant for a number of ecological and economic factors, such as reduction the impact on the environment, increase energy and resource efficiency of the production process, insurance the possibility of commercialization of the waste at the newly created market consumption, the organization of commercial production in the operation of waste recovery.

Industrial production processing of polymeric materials (for example, the production of car tires) are associated with the formation of a very specific production waste in the form of the rubber polymer edged cord or steel cord. During the formation of the tire on the molds with the layering of flat rubberized cord is formed with their shotguns dimensional parameter and configuration unsuitable for the production of primary products. At the same time, cutting rubberized cord retain their physical and mechanical properties that are typical of the raw material in the production of rubber-technical products, which determined the urgency of solving the problem of the process of development of this type of waste recovery with purchase of the raw material specified quality index.

The result of research and development in the development process waste tire production (JSC "Nizhnekamskshina") defines an innovative technology to produce a new type of composite material - fiber-filled rubber compound (FFRC), which is a composite raw material for the production of rubber products with high physical and mechanical operation.

The proposing technology of waste products recovery through the use of rubberized cord edges has the following advantages:

1. At present, wastes formed during the production of tires as rubberized cord edges are giving to combustion in thermoovens, which is associated with the costs to capture pollutants of second and

third categories of danger and payments for emissions. There is a significant turnover of material resources.

2. In the case of giving these wastes to consumers the costs of waste management are transformed into profits of the enterprise.
3. The use of waste recycling within the company by opening a small enterprise for the production of new material (FFRC) determines the conditions to expand the market of products, lays the technological possibilities of manufacture of alternative kinds of rubber goods. In particular, it is possible to ensure that import substitution in the oil producing enterprises (OJSC "Tatneft") through the production of high-quality V-belts for oil rocking chairs on the basis of a new type of raw materials - composite fiber-filled rubber compound.

An innovative approach to technology waste tire production to produce a fiber-filled composite polymeric materials defined the development and manufacture of a new patent-protected equipment for dispersion and mixing extruder protected by patent "Device for the preparation of fiber-filled composite polymeric materials" (patent №73269).

The design of the extruder allows grinding rubberized tire cord edges to the desired structural dispersion cords, promotion of small segment to nanostructures and further mixing with a rubber component to obtain a homogeneous composite rubber compound with fine-fiber filled nylon.

Designed extrusion process of obtaining a new type of composite material by utilization of secondary resources of tire production - recovery shotguns rubberized cord realized by conducting design work with the production of a pilot plant - precipitation mixing extruder DSE-1.

Research of performance characteristics of rubber products manufactured on the basis of the produced composite rubber mixture FFRC has been conducted at the Laboratory of elastomers KNIITU with the assessment of the curing properties on the «Reomtr-100S». Compared with caoutchouc rubber mixtures of general purpose rubbers qualitative characteristics on the basis of curing properties of FFRC showed significant positive differences. In particular: conventional tensile strength reaches 22.8 - 24 MPa; elongation 10%; tear resistance reaches 125 - 149%; the Shore hardness value of 92 - 95; rebound resilience of 35 - 37%; loss on attrition is reduced to (0.03 - 0.04)10^{-3} kg.

The tests found that rubber-based FFRC have high wear resistance, high adhesion to the rubber-cord, which ensures a long life design.

Expansion (diversification) Nomenclature RTP based FFRC involves: the ability to produce a wide range of V-belts, flat (ribbed) belts for trucks, agricultural machines, machinery equipment; increased resource exploitation RTP; the ability to manufacture a wide series of V-belts for the oil fields, the engines of automobiles "KAMAZ"; production of high quality components for industrial rubber goods for transport machinery (dampers, pulleys, rollers, scrapers snowplows); Issue wear-resistant conveyor belt; the ability to produce under-rail pads with high resistance to bending and impact strength, able to serve on construction of high-speed railway lines.

The following particular specific properties are achieved with the introduction of high-strength fiber materials: for carbon fiber and kevlar - impact strength, armor; to asbestos fibers - brake wear.

In the direction of applying the results of research the technological capabilities of industrial waste recycling are established on the large-scale production of polymer products, both directly in the production of composite materials, and the resulting application of a new type of raw material in the manufacture of rubber goods (3).

As a result of research on the topic, it was designed and manufactured mold for the production of V-belts series C-4000 for oil rocking chairs. A pilot batch of V-belts C-4000 were field tested in the oil fields of OJSC "Tatneft". V-belts made by the proposed technology have worked 9,000 - 12,000 operating hours (setting of 2 belts), while the belts of the company "Otibelt" (Germany) have the resources of 5000 operating hours with the setting of 3 belts.

In addition to the presented results it should be noted that the scrapers for conveyor harvester "Don" were made from the resulting FFRC which worked for 3 seasons. Whereas, under similar conditions as the traditional rubber scrapers ensure resource for one season.

The proposed technology is concluding the following advantages:

- The majority of production at the plant for the production of industrial rubber goods, including tire plants in Russia (in the cities as Nizhnekamsk, Voronezh, Omsk, Kirov, etc.) and other countries, the solution of problems associated with waste disposal and recycling of secondary material resources. The proposed technology solves these problems;
- In the innovative engineering technology it solves the problem of providing new types of raw materials, which defines the production of a new product with high performance - composite materials;
- By developed technology and advanced equipment it enables the organization of industrial production with innovative technology of manufacture of transmission belts having a world-class quality indicators and a wide area of consumption - oil production, automotive, agricultural mechanical engineering, machine tool equipment.

The proposed technology for their promising direction is determined by the possibility of 100% recycling - as secondary resources for tire factories of the Russian Federation to ensure the production of many kinds of high quality RTP with specific indicators.

Currently, in KFU it is going the research on innovative technology of utilization of metal-rubberized cord edges (MRCE). RTP obtained by MRCE have required level of performance of elasticity, high impact strength, abrasion resistance, armor protection to be penetrated, which are opening up opportunities for the diversification of technology in the fields of mechanical engineering, road construction, military equipment.

As part of this research project, the pilot batch of scrapers for snow plows made from MRCE have passed field trials in winter 2014/15, and will be continued in 2016.

References

[1] Husainov A D, Azimov Y I 2010 Repeated vulcanizators on the base of rubber fibrous materials *Vestnik KTU* vol 9 pp 356-360

[2] A G Emelyanov 2004 Ecological bases of nature management (M .: Publishing House Academia) 304 p.

[3] Gilmanshin I R, Ferenets A V, Azimov Yu I, Galeeva A I, Gilmanshina S I 2015 Innovative technologies of waste recycling with production of high performance products *IOP Conference Series: Materials Science and Engineering* Vol 86 pp.12014-16

[4] Gilmanshin I R, Kashapov N F, Azimov Y I, Gimanshina S I, Ganeeva D A, Valishev R D 2015 Utilization of biogas solid waste landfills by building power complex on the basis of alternative energy installations *Kazanskii economicheskii vestnik* Vol 2 (16) pp 41-44

[5] Dautov G, Dautov I, Fayrushin I and Kashapov N 2013 *J. Phys.: Conf. Ser.* **479** 012001

[6] Kashapov L, Kashapov N and Kashapov R 2013 *J. Phys.: Conf. Ser.* **479** 012005

Balancing of the anthropomorphous robot walking

V M Devaev, D V Nikitina, A Y Fadeev

Kazan (Volga region) Federal University
18, ul. Kremlevskaya, 420008, Kazan, Russia

vdevaev@yandex.ru, daria.92rus@gmail.com, andrjuhafad@ya.ru

Abstract
Anthropomorphic robots are designed a human environment operates: buildings and structures, cabs and etc. The movement of these robots is carried out by walking which provides high throughput to overcome natural and manmade obstacles. The article presents some algorithm results for dynamic walking on the anthropomorphic robot AR601 example.
The work is performed according to the Russian Government Program of Competitive Growth of Kazan Federal University.

Modern requirements for robotics working in conditions of accidents and emergencies are very complex, so the anthropomorphic robots have to work in human existence environment with the tools and apparatus people adapted for [1]. DARPA projects of the US Ministry of Defense announced a contest (DAPRA robotics challenge) of the best robot design are able to move cross-country, climb stairs, use regular electric tools and drive vehicles. Recent developments in the nuclear power plant "Fukushima" has shown the necessity to use robots for disaster. This defines the high requirements for stability of walking anthropomorphic robots and anthropomorphic with cargo [2].

Creating a walking robot - a task required solving many interrelated problems including the problem of motion simulation. Walking anthropomorphic mechanisms have a large number of freedom degrees and their motion described by complex equations [3]. For development of algorithms and software traffic management, we offered to consider complex controlling robot walking as a set of simple patterns or patterns of movement developed by analogy with conditioned and unconditioned reflexes like implemented steps of human. In the paper presents results of development of the not-supporting leg pattern moving for a predetermined path and stabilize the robot actuators of supporting leg.

To achieve this goal highlighted two problems. The first subtask - moving the not-supported feet of anthropomorphic robot of a parabolic trajectory by a predetermined distance. We assume that the motion of a parabolic trajectory, or "stone's throw" is more keep in terms of energy [4]. The second subtask - providing static stabilization of the robot while standing and the committed step. We consider a plane-parallel movement of the legs in the robot plane of symmetry.

Introduce the inertial reference system: the x-axis - is parallel to the Earth's surface in the direction of movement of the robot, Oy - perpendicular to the surface of movement and the Ox axis vertically (Fig. 1). Position the hip (OA) and tibia (AB) about plane xOy defined angles α and β. From foot

IOP Conf. Series: Materials Science and Engineering **134** (2016) 012004 doi:10.1088/1757-899X/134/1/012004

coordinates (p.B) we need to find the angles α and β at each time.

Moving a robot feet is made of a parabola described by the equation:

$$y = \frac{-4h}{L^2} * x^2 + \frac{4h}{L} * x (1)$$

where, L - step length, h - step height.

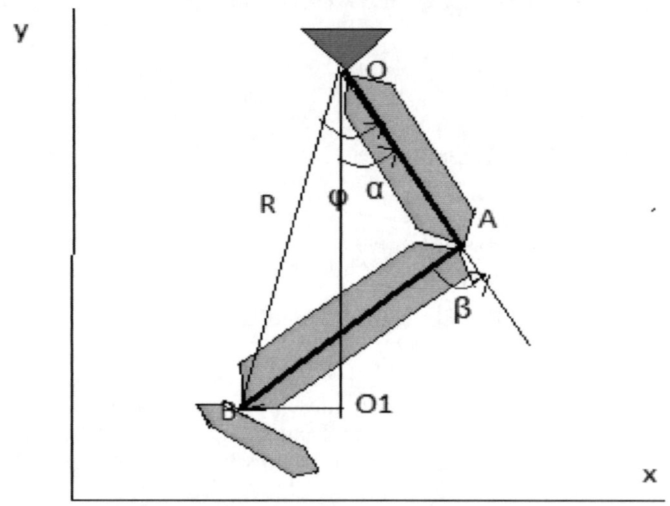

Fig.1. Robot leg geometric scheme

The distance between the hip and the knee, as well as between the knee and shin are (OA = AB = l2 = l1 = 280 mm);
From geometrical scheme in Fig. 1, we find:

$$R = \sqrt{(X_A - X_O)^2 + (Y_A - y_O)^2} \quad (2)$$

$$\phi = \arccos\left(\frac{l2^2 - l1^2 + R^2}{2R * l2}\right) \quad (3)$$

Then find the angles value:

$$\alpha = \varphi - arctg\left(\frac{OO_1}{O_1C}\right) \quad (4)$$

$$\beta = 180 - 180 - 2*\varphi = -2 * \varphi \quad (5)$$

The path of the not-supported leg were built in Octave software [5].

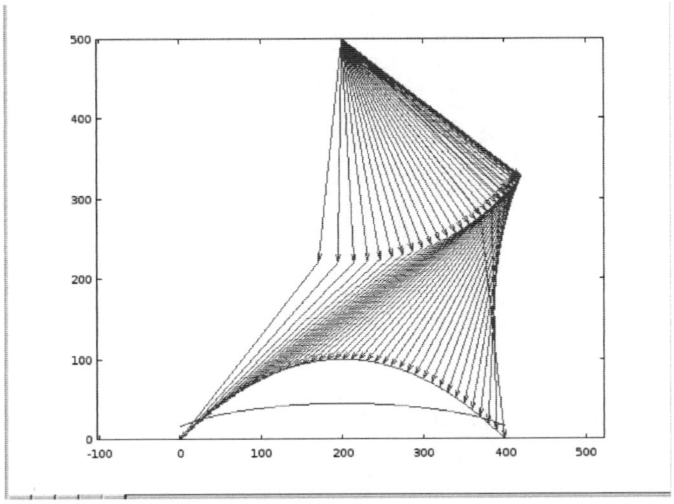

Fig.2. Moving not-supported leg on a parabolic trajectory.

Software was written for working of this movement pattern on the AR601 [6]. According to the given values of the length stride (L), the step height (h) and the time step of one feet (T), built the path of the not-supported robot leg, one of which is presented below:

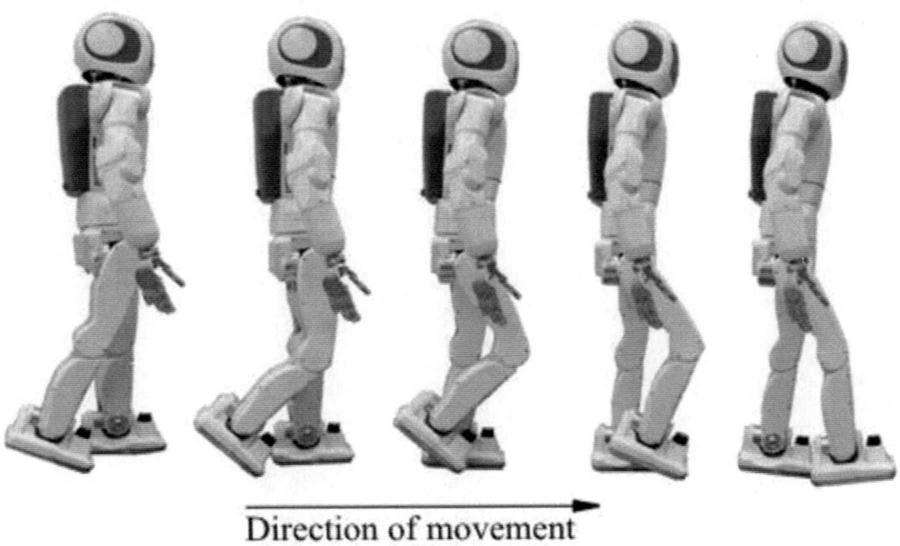

Fig.3. Moving not-supported robot leg on a parabolic trajectory.

To measure the deviation of the anthropomorphic robot AR 601E of the vertical use forces and moments strain gauges placed near the feet joints. It species are represented in Fig.4.

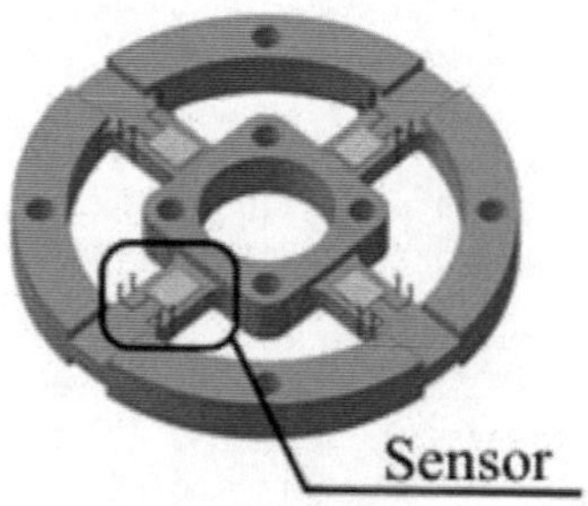

Fig. 4. Forces and moments sensor.

We can detect inclination of the robot by calculating the difference between the values of the pair of opposing sensors. For that reason drawn up a calibration table in which a robot angle corresponds to a particular value of the difference foot sensors. It is also define limits of the robot deviation to ensure stability, it was 8 degrees forward fand, 4 degrees backward. Software was written at Python that in depends of the difference sensor value generates a drive control located at the feet and return robot to an upright position after deflection from the vertical longitudinal plane.There is a video showing the performance of the developed method of the robot stabilization.

References

[1] Hirose M, Ogawa K 2007 Honda Humanoid Robots Development. *Philosophical Transactions of the Royal Society A* **365** 11-19
[2] Defense Advanced Research Projects Agency (DARPA)
[3] URL http://www.darpa.mil/about-us/about-darpa
[4] A V Voronov 2004 Simulation biomechanical modeling as a method of studying motor of human actions, Scientific and Technical Journal **2** 22-26, 39-40 http://lib.sportedu.ru/Press/TPFK/2004N2/p22-26,39 -40.htm
[5] V M Devaev 2015 Minimax method for estimating the parameters of motion robotic systems for multiple measurements of small precision. XI All-Russian Congress of the fundamental problems of Theoretical and Applied Mechanics 1148-49
[6] GNU Octave URL http://www.gnu.org/software/octave/
[7] Android Technics URL http://npo-at.com/products/ar-600e/

Mathematical modeling of jet flow around of bodies when applying polymer powder coatings

V L Fedyaev[1,2], E R Galimov[2], I R Gimranov[3], M S Takhaviev[3] and I V Morenko[1]

[1] Institute of Mechanics and Engineering, Kazan Science Center, Russian Academy of Sciences, Kazan, Russian Federation
[2] Kazan National Research Technical University named after A N Tupolev, Kazan, Russian Federation
[3] Gazprom Transgaz Kazan, Russian Federation

E-mail: morenko@imm.knc.ru

Abstract. Mathematical modeling of processes occurring during the spraying of polymer powder compositions on the surface of parts and products is considered. Relations describing the flow around processed bodies, movement of the powder particles in a stream of the carrying medium, dynamics of their deposition on the body surface are presented.

1. Introduction

At the present time, coatings based on the polymer powder compositions (PPC) [1] are becoming more common in many industries, construction and transport. Promising methods of PPC spraying are jet methods such as flameless spraying on a heated surface of parts and products, hot spraying, spraying under the influence of centrifugal forces in an electrostatic field. Internal gas-dynamic, thermal problems in the cavities of spraying guns [2], other spraying devices and external problems when considering jet nonisothermal inleakage of the carrying medium (air, combustion products) containing the powder particles on the surface of parts and products are distinguished at mathematical modeling of processes occurring in the implementation of the named methods. It should be noted that in these problems apart from the dynamics of gas and particles the heat exchange of powder particles with the carrying medium, thermal behavior of the powder coating material layer and in some cases the heat exchange of a jet with the environment are of great importance. Similar problems are encountered at jet painting, abrasive machining of parts and products, work of ventilation systems of industrial and residential premises, reversing devices, when studying the processes occurring during the takeoff of the aerospace objects and in other cases [3-6].

2. Relations describing the flow around processed bodies, movement of the powder particles, dynamics of their deposition on the body surface

Models of an ideal either viscous compressible or incompressible fluid can be used for a mathematical description of the medium movement when spraying the polymer powder compositions on parts and products according to the classification of jet streams [7]. In this case, the medium can be considered as a homogeneous fluid. However, in the considered processes it definitely represents a multiphase gas suspension or depending on the particles sizes the aerosol. Wherein disperse particles may or may not have an electric charge. Furthermore, as mentioned above, it is necessary to take into consideration the interaction of the jet with the environment. A jet can spread in a stationary or moving medium in the absence or presence of limitations (straitened jets). When the jet flows around parts and products the latter also represent limitations moreover they can be of very different shapes, sizes, they can move, rotate relatively to the source of the jet or be stationary. Finally, plane, axisymmetric, circular, fan, nonswirled or swirled jet can be identified by the shape and nature of medium movement in the jet.

In this work it is assumed that the jet fluid is ideal, incompressible and contains small sized uncharged particles of a polymer powder. There are no limitations of a jet flow with the exception of a streamlined body. Thermal interaction of a jet with environment and particles with the carrying medium is disregarded. Since in practice mainly circular or slotted nozzles are used when spraying only plane and axisymmetric gas powder jets are considered below.

For a plane-parallel fluid flow solutions are obtained as a rule by the conformal mapping method, for example, the problems of jet inleakage on the plate of finite length, semi-infinite and infinitely stretched plate [8-10 and others]. Peculiarities of the jet flow around the bodies of arbitrary cross-section, circular cylinder, in particular, rotating one are discussed in [11]. Free jet, zone of flow reversal and jets spreading over the surface, probably, separating from it can be marked out in the flow region when a fluid or gas jet flows at a moderate speed on the body. Specificity of studies of jet dynamics, movement of powder particles in the flow when spraying is that from the practical point of view the greatest interest represents the zone of flow reversal in the vicinity of the body front point where particles deposit on the surface.

In the case when the width of the jet cross-section near the body is greater than its transverse size it can be expected that the jet flow in the reversal zone differs only slightly from the flow in the same region when unbounded stream flows around a body. Here, solutions to the problem of an unbounded stream flowing around a body can be used in the said zone to assess the behavior of the powder particles. Thus, for flow around a stationary circular cylinder of radius a velocity vector components of a fluid in the cylindrical coordinate system (z,r,φ) are:

$$v_r = v_0 \cos\varphi\left(1-a^2/r^2\right),\quad v_\varphi = -v_0 \sin\varphi\left(1+a^2/r^2\right). \tag{1}$$

Here v_0 – fluid velocity at a distance from the body directed along the r-axis.

If the cylinder rotates around the Oz-axis counterclockwise with an angular velocity ω then flow function is:

$$\psi = v_0 r \sin\varphi\left(1-a^2/r^2\right) - \omega a^2 \ln r, \tag{2}$$

velocities $v_r = \partial\psi/(r\partial\varphi)$, $v_\varphi = \partial\psi/\partial r$.

Further let us consider the spatial flow around a body. If the body is a sphere of radius a, the flow is axisymmetric and the symmetry axis coincides with the Oz-axis of cylindrical coordinate system (z,r,φ) with the origin at the center of the sphere then according to [12] flow function is:

$$\psi = -0.5 v_0 r^2\left(1-\left(a^2/\sqrt{r^2+z^2}\right)^3\right), \tag{3}$$

velocities $v_r = \partial\psi/(r\partial z)$, $v_z = -\partial\psi/(r\partial r)$, $v_\varphi = 0$.

In the case of axisymmetric flow around a semi-infinite body of rotation the flow function is

$$\psi = -0.5 v_0\left(r^2 - 0.5 r_\infty^2\left(1+z/\sqrt{r^2+z^2}\right)\right), \tag{4}$$

where r_∞ – radius of the body cross-section removed far from the front point located at a distance of $0.5r_\infty$ from the origin.

In addition to the mentioned above cylindrical coordinates we introduce Cartesian coordinates (x, y, z) with the origin O. The Ox-axis is directed along the velocity vector of the oncoming flow. At a distance from the surface of the subject body, upstream, in the transition region from the zone of a free jet to the zone of flow reversal we single out section S_0 perpendicular to the velocity vector of the oncoming flow. We assume that in this section polymer powder particles are thrown into the flow. Confining ourselves to the case of two-dimensional flow of the medium (plane-parallel or axisymmetric flow) the initial coordinates of the particles are denoted as x_{p0}, y_{p0}, and velocity vector components as u_{p0}, v_{p0}. For simplicity we assume that all powder particles are spherical, over time their radius r_p does not change. Possible rotation of the particles around its own axis is neglected, particles do not stick together and do not break up.

Following the Lagrangian continual approach, neglecting the forces of gravity, added masses and Bosse, to describe the motion of powder particles in a fluid we use the following relations:

$$\frac{du_p}{d\tau} = R_\mu\left(u - u_p\right), \quad \frac{dv_p}{d\tau} = R_\mu\left(v - v_p\right). \tag{5}$$

Here τ – time, $u_p = dx_p(\tau)/d\tau$, $v_p = dy_p(\tau)/d\tau$ – velocity components, $x_p = x_p(\tau) = x_{p0} + f_p(\tau)$, $y_p = y_p(\tau) = y_{p0} + g_p(\tau)$ – law of the particles motion; $R_\mu = k_c r_p \mu / m_p$, m_p – particle mass, μ – coefficient of fluid viscosity. We assume that the hydrodynamic forces acting on the particles are proportional to the difference of velocities of the fluid and particles, k_c – proportionality coefficient (k_c=19÷25).

Considering that in equations (5) velocity vector components of a fluid which, in particular, can be determined by the formulas (1-4) depend on time $u = u\left(x_p(\tau), y_p(\tau)\right) = U(\tau)$, $v = v\left(x_p(\tau), y_p(\tau)\right) = V(\tau)$ we present them in the form: $U = a_0 + a_1\tau$, $V = b_0 + b_1\tau$. Substituting U, V in (5) and integrating these equations we find

$$u_p = u_{p0}\exp\left(-R_\mu\tau\right) + \left(a_0 - a_1/R_\mu\right) + a_1\tau; \tag{6}$$

$$v_p = v_{p0}\exp\left(-R_\mu\tau\right) + \left(b_0 - b_1/R_\mu\right) + b_1\tau.$$

Hence the particles coordinates

$$\begin{aligned} x_p &= x_{p0} - \left(u_{p0}/R_\mu\right)\exp\left(-R_\mu\tau\right) + \left(a_0 - a_1/R_\mu\right)\tau + 0.5a_1\tau^2; \\ y_p &= y_{p0} - \left(v_{p0}/R_\mu\right)\exp\left(-R_\mu\tau\right) + \left(b_0 - b_1/R_\mu\right)\tau + 0.5b_1\tau^2. \end{aligned} \tag{7}$$

Under condition that $R_\mu\tau < 1$ these dependencies can be simplified:

$$x_p \approx A_0 + A_1\tau + A_2\tau^2; \quad y_p \approx B_0 + B_1\tau + B_2\tau^2, \tag{8}$$

where coefficients $A_0 = x_{p0} - u_{p0}/R_\mu$, $A_1 = u_{p0} + a_0 - a_1/R_\mu$, $A_2 = 0.5a_1$; $B_0 = y_{p0} - v_{p0}/R_\mu$, $B_1 = v_{p0} + b_0 - b_1/R_\mu$, $B_2 = 0.5b_1$.

Relations (7, 8) are equations of particles trajectories in the parametric form. Using them it is possible to find the intersection points of these trajectories with the streamlined body surface, estimate the mass distribution of the powder on its surface.

In this paper we propose a slightly different approach. It is assumed that the processed part of the surface of a detail, product is convex there are points M_1, M_2 on it at a maximum distance from the Ox-axis their coordinates are (x_{m1}, y_{m1}), (x_{m2}, y_{m2}), respectively. Substituting the value x_{m1} instead of x_p in the first of equations (8) we find the time $\tau = \tau_{m1}$, during which the particle with coordinates $(x_{p0}, y_{m1} = y_p(\tau_{m1}))$ in section S_0 at the initial time reaches the point M_1. Similarly, the initial position of the particle falling into the point M_2 on the body surface is determined. It is obvious that all particles in section S_0 with coordinates $(x_{p0}, y_{m2} < y_p < y_{m1})$ can reach the body surface and stay on it.

Further, it is convenient to break the arc part of the body surface between M_1 and M_2 by points K_i on n equal parts of length l_0 $(i = 1, n-1)$, find the coordinates y_{p0i} of the initial position of the powder particles which have fallen into points K_i.

It is easy to make sure that mass flow rate of the particles material passing through the part $(\Delta_{pi} \cdot 1)$ of the section S_0 can be estimated by the dependency:

$$G_i \approx \rho \bar{\varepsilon}_{pi} \bar{u}_i (\Delta_{pi} \cdot 1). \quad (9)$$

Here, ρ – material density of the polymer powder particles, $\bar{u}_i$ – normal to the section S_0 average velocity of the particles which are in the part $\Delta_{pi} = y_{p0(i+1)} - y_{p0i}$ at the initial time, $\bar{\varepsilon}_{pi}$ – average volume concentration of particles of a gas suspension flowing through this part.

For the time of spraying $\tau_H > (\tau_{m1}, \tau_{m2})$ mass of the powder that has entered the part of a body surface between the points K_i and K_{i+1} is $\tau_H G_i$. Accordingly, under the condition of its uniform distribution over the said area thickness of the sprayed layer is

$$h_i \approx k \bar{\varepsilon}_{pi} \bar{u}_i \tau_H (\rho \Delta_{pi} / (\rho_i l_0)), \quad (10)$$

where ρ_i – medium density of the powder layer which depends on the falling speed of particles on the surface, their size, degree of fusion and other factors; k – correction coefficient that can be used to take into account the powder movement along the body surface, its blowing.

3. Conclusions

In conclusion we note that peculiarities of flow around bodies by carrying medium, influence of such parameters as speed of oncoming flow v_0, angular velocity ω when the body rotates, characteristic dimension a, body shape (see (1)-(4)) as well as the value R_μ depending on the particles radius r_p, their mass m_p, medium viscosity μ, finally, distance of the initial section S_0 from the body, position of the reference part K_i, K_{i+1} on its surface are taken into account in the expression for determining the thickness of the sprayed layer h_i (10) indirectly through the average speed $\bar{u}_i$.

References

[1] Galimov E R, Zverev E V, Tukbaev E E, Galimova N Ya, Kuryntsev S V and Mukhin A M 2012 *Polymer powder coatings for special purposes* (Kazan: Ofset Servis)
[2] Fedyaev V L, Morenko I V, Siraev A R, Galimov E R, Gimranov I R, Fazlyev L R and Takhaviyev M S 2015 *IOP Conf. Series: Materials Science and Engineering* **86** 012033
[3] Abramovich G N, Girshovich T A, Krasheninnikov S Yu, Sekundov A N and Smirnova I P 1984 *Theory of turbulent jets* (Moscow: Nauka)

IOP Conf. Series: Materials Science and Engineering **134** (2016) 012005 doi:10.1088/1757-899X/134/1/012005

[4] Kashapov L N, Kashapov N F and Kashapov R N 2013 *J. Phys.: Conf. Ser.* **479** 012003
[5] Dautov G, Fayrushin I, Kashapov N F and Dautov I 2014 *J. Phys.: Conf. Ser.* **567** 012008
[6] Snigerev B A 2015 *J. Engineering Physics and Thermophysics* **88**(1) 230
[7] Vulis L A and Kashkarov V P 1984 *Theory of viscous fluid jets* (Moscow: Nauka)
[8] Bai Shi-i 1960 *Theory of jets* (Moscow: Fizmatgiz)
[9] Birkgof G and Zarantonello E 1964 *Jets, wakes and caverns* (Moscow: Mir)
[10] Gurevich M I 1961 *Theory of ideal fluid jets* (Moscow: Fizmatgiz)
[11] Lavrentyev M A and Shabat B V 1973 *Problems of hydrodynamics and their mathematical models* (Moscow: Nauka)
[12] Kochin N E, Kibel I A and Roze N V 1963 *Theoretical hydrodynamics* (Moscow: Fizmatgiz)

IOP Conf. Series: Materials Science and Engineering **134** (2016) 012006 doi:10.1088/1757-899X/134/1/012006

The formation of students' engineering thinking as a way to create new techniques, technologies, materials

Iskander Gilmanshin[1], Suriya Gilmanshina[1]

[1] Kazan (Volga region) Federal University, 420008,
Russia, Kazan, Kremlevskaya street, 18

e-mail: is-er@yandex.ru

Abstract. Engineering thinking is regarded as the quality of the person, which is stimulating the human need for the creation of new techniques, technologies and materials. Applications in the study of competence approach allows us to consider a professional thinking as one of the core competencies required for successful engineer innovations in mechanical engineering. The author's definition of professional engineering thinking is presented. The ways of its formation at students of technical fields enrolled in university courses are illustrated

In modern conditions of rapid aging of knowledge engineer's professional mobility is defined by personal qualities. An essential quality of personality, which is stimulating his need for the creation of new techniques, technologies, materials could be professional engineering thinking. Competence approach, on which modern educational standards are based, it can be considered that professional thinking is one of the core competencies required for successful engineer innovative developments in the field of mechanical engineering.

Professional thinking is generic with respect to the whole group of selected forms of thinking in the literature related to a particular profession or group of professions. Its specificity is manifested in the activities and tasks determined by the content, problem situations that can be solved on a professional stage career path. This is the content side of professional thinking and engineering as well.

For our research of interest to the following understanding of the essence of engineering thinking. Engineering thinking is a specific mental activity, in which the reflection and creative transformation in the human psyche phenomena of engineering design, construction, creation and development of new techniques, technologies, materials such as social functions, as well as a subjective construction of the engineering process in accordance with the social objectives of vocational education .

Features of professional engineering thinking caused by the specifics of the functions of this thinking. When performing certain functions of an engineer thinking it is intuitive and practical, performing other - research.

Formation of the students enrolled in technical fields, engineering thinking in the learning process in the classical university is a lengthy process of gradual implementation of a number of interrelated milestones. Today within the competence approach invariant system of intermediate objectives is to require a gradual transition of mental activity of the future engineers with the reproductive level to the creative activity of the creation and development of new techniques, technologies and materials. Such a transition requires familiarizing students to the analysis of contradictions and formulation of problems, hypotheses, their justification and proof in explaining the technical processes and phenomena.

This multifunctional and gradual process can be carried out as follows. When using a university professor explanatory and illustrative method, students follow a course of reasoning teacher. The emphasis is on the logical structure of explanation of the essence of the process or phenomenon. Further, the application of the problem of presentation of the lecture material, students are involved in the formulation of problems at the beginning of lessons, as well as finding solutions to the problems with the help of a teacher using appropriate methods of scientific reasoning (analogy, inductive and deductive assumptions, modeling, thought experiment [6]). When using more complex partial retrieval method (heuristic conversation) students together with lecturer analyze and generalize the experimental data, formulate hypotheses, conjectures to solve the problem, explain and prove the hypothesis of theoretical and empirically. The highest level of search can be achieved independence of students in the classroom, carried out by research.

As the integration processes in vocational education has always attracted the attention of teachers and researchers [2, 7], separately it should be noted working out special methods of solving experimental and computational engineering problems on the basis of interdisciplinary connections, systematic implementation of which learning is the basis for the creation of integrated educational and informative technical problems . On the theoretical and practical lessons in the process of solving these problems should also apply various methods of scientific knowledge with the gradual nature of the complication of tasks to be done by students. Otherwise, organize mental activity of students in the classroom (for example, energy-saving technologies) should be such that in the process of analyzing the students contradictions create a practical problem situation, they would arise subjective state of intellectual puzzlement and desire to get out of this state. We believe that the problem situation as a certain structural component of creative thinking is an efficient unit of analysis is the level of professional thinking of students of Engineering.

So, a special work on formation of students' engineering thinking, involves a problem of lectures in which students get acquainted with the methods of analysis of problem situations encountered in practice, setting the logic of engineering problems and ways of solving them. In order for students to participate in solving the problems of the lecture, the teacher is required at the same time outlining the lecture material to solve a number of problems. First, present the facts without a detailed explanation in logical terms "reasoning" presentation, which stimulates mental activity of students. Second, to control cognitive activity of students during a lecture. Third, clarify the ways in which the problem is solved lecture. There is a marked contradiction between the requirements for the organization of problem-based learning in a lecture and the real possibilities of their implementation in a lecture at the university. To overcome it requires a specific approach to the development of problem lecture. It should take into account four aspects: general didactic, informative, teaching and learning and educational.

Consider obschedidakticheskie problematic aspect of the lecture. By the nature of the interconnected activity of the teacher and students of a problematic presentation takes into consideration characteristics of productive thinking of students [1]: the creation at the beginning of the lecture the problem situation, based on its analysis and identify contradictions formulate problematic issues to find ways to solve the problem due to the reformulation and hypotheses, choice and prove the most faithful to the hypothesis about the way to solve the problem. Taking into account these features of productive thinking in presenting the material and activation using heuristic methods of mental activity of students, aimed at addressing the lecture provides a problematic approach to the presentation of the lecture material.

It is necessary to distinguish between substantive (analysis of one of the problems of science and practice, such as the problem of obtaining products with high performance [4], the formation of a regional system engineering [3]) and methodological (application of the method of presentation of the problem) side problem lecture.

Mental activity stimulates the students as the use in the educational process of design training (more in [5]) and the method of hypotheses. This requires a combination of presentation lecture topics with logical assignments to students. These tasks can be called didactic methods of control anticipatory cognitive activity. For example: 1) The teacher presents factual material, students are invited to put forward in the course of the lecture series of particular assumptions that are integrable in the final stage in the general hypothesis. Last checked generalization that makes the teacher at the end of the lecture; 2) the teacher at the beginning of the lecture the general hypothesis put forward on this problem, in the course of the lecture, students supplement and clarify the hypothesis; 3) at the beginning of the lecture, after the formulation of the problem formulated two conflicting hypotheses, reflecting the contradictory nature of the process or phenomenon being studied. Next, the students followed the course of the lecture, dialectically synthesized these hypothetical statements and make meaningful generalizations.

Thus, the communion of future engineers to creative mental activity on creation and development of new techniques, technologies and materials based on the realities of today through the analysis of contradictions, and posing problems, hypotheses, their justification and proof.

References

[1] Vilkeev DV 2001 Educational Psychology: lectures (Kazan: KSPU) 262 p

[2] Integration processes in modern vocational education: collective monograph Ed. GV Mukhametzyanova 2013 (Kazan: Publishing house "Printing service XXI century") 356 p

[3] Gil'manshin I R, Kashapov N F 2014 Energy service contracts in regional engineering center for small and medium businesses *IOP Conference Series: Materials Science and Engineering* (Bristol-UK) Vol 69

[4] Gilmanshin I R, Ferenets A V, Azimov Yu I, Galeeva A I, Gilmanshina S I 2015 Innovative technologies of waste recycling with production of high performance products *IOP Conference Series: Materials Science and Engineering* Vol 86 pp.12014-16

[5] Gilmanshina S I, Gilmanshin IR 2015 Building axiological competence of graduate students by means of project-based learning *IOP Conference Series: Materials Science and Engineering*, Vol 86 pp12029-32

[6] Gilmanshina S I, Sagitova RN and et 2015 Professional thinking formation features of prospective natural science teachers relying on the competence-based approach *Review of European Studies* Vol 7 No 3 pp 341-349

[7] Sibgatova K I, Gilmanshina S I, Khalikova F D, Gilmanshin I R, Akchurina I R and et 2015 Peculiarities of pupils and vocational college students' career guidance modeling in the integrated system "School – College – Enterprise" Asian Social Science Vol 11 No 1 pp 386-391

[8] Dautov G, Fayrushin I, Kashapov N and Dautov I 2014 *J. Phys.: Conf. Ser.* **567** 012008

[9] Fayrushin I, Kashapov N and Dautov I 2014 *J. Phys.: Conf. Ser.* **567** 012009

IOP Conf. Series: Materials Science and Engineering **134** (2016) 012007 doi:10.1088/1757-899X/134/1/012007

Landfill energy complex based on the renewable energy installations

Iskander Gilmanshin[1], Nail Kashapov[1], Suriya Gilmanshina[1], Asiya Galeeva[2]

[1] Kazan (Volga region) Federal University, 420001, Russia, Kazan, Kremlevskaya street, 18
[2] Power Engineering University, 420066, Kazan, Krasnoselskaya str., 51

e-mail: is-er@yandex.ru

Abstract
The article presents the analysis of standard approaches to degassing of landfills. The need of comprehensive work on the degassing of the landfill body is identified. The author's task decomposition of the landfill degassing is formulated. The analysis of existing methods of work on degassing of landfills is presented. The author's approach is including implements of series of parallel studies in the framework of achieving a common goal to reduce the anthropogenic pressure on the ecosystem of the region due to the need for disposal of solid waste. An action plan for the development of the target problem-oriented management techniques of the landfill with the following development of the effective energy complex is formulated.

To the issue of landfill gas utilization has traditionally being paid great attention in Europe, Latin and North America. However, this issue is relevant for the landfills of the Russian Federation as well. The need for complex work of degassing of the landfill due to a number of factors: the withdrawal of the effective turnover of large areas in the area of the city; the effect of methane on global climate change is estimated to be 23-25 times higher than the potential impact of carbon dioxide; allocation of landfill gas creates the preconditions to spontaneous combustion of deep layers of the landfill; utilization of the landfill gas can generate additional electrical and thermal power and allow to reduce human pressure on the ecosystem of the region, and in some cases extend the life of landfill.

The severity and urgency of these problems are increasing day by day. The intensive development of the economy increases the human impact on the environment. Thus, research in the field of construction of a modern, efficient energy complex based on a closed or operated landfill is an urgent task.

After researching the task, the several areas of research are determined: the study of the morphological composition of the landfill; modeling of landfill internal processes lifecycle; analysis of the energy potential of the landfill; the development of problem-oriented methodology of the landfill degassing; system analysis of the process of processing, recycling and disposal of solid waste, development of effective management methodology of landfills and building an effective energy complex on the basis of alternative energy systems.

Analysis of the literature has showed that there are a number of developing and applying various technologies of degassing technologies have place today [4, 5, 6, 7]. However, conducted researches have very scattered narrowly focused work. The effective problem solving possible with a systematic approach. It is necessary to consider the landfill as a holistic energy complex such as power plex of industrial enterprises [9]. That will have the further attractiveness of the project to investors. [10] Due to the information above, we propose an approach that implements a series of parallel studies in the task of the work to achieve a common goal: reduce the anthropogenic load on the ecosystem of the region due to the need for disposal of solid waste.

The algorithm of research towards the study of the morphological composition of the landfill.

The most landfills of Russian Federation have formed without regard for the specific use as platforms deployment of alternative energy systems. Moreover morphological structure of exported waste has undergone significant changes and will change further. [11]

In such circumstances, it is important to define the boundary conditions of the functioning of the energy complex in conjunction with morphological composition and distribution of the landfill body. Analysis of the grounds revealed insufficient monitoring of the morphological composition of the waste is actually: a lack of strategic planning in the formation of the landfill; absence or chaotic actions of the administration of waste sorting and selection of commercially attractive materials.

It should be noted that in connection with the proceeding in the landfill process of natural decomposition of organic compounds in the morphological structure of its quantitative and qualitative indicators have variable numbers. The energy potential is directly related to the intensity and the concentration of the decomposition of organic compounds in the body of the landfill.

According to the statistics, solid waste consists of 55 - 70% organic compounds. The process of decomposition of this waste takes place in many ways identical to the process of fermentation of highly contaminated sewage sludge or sewage treatment facilities. As a result biogas is produced with combustion heat 18900 - 25100 kJ / m3 (4,500 - 6,000 kcal / m3). Average composition of biogas is the following: methane - 50-65%, carbon dioxide - 30 - 45%, hydrogen sulfide - 0.2-0.8%, nitrogen, oxygen, hydrogen, - 1-2% and aromatic hydrocarbons – up to 1% in the conditions of the average moisture content of biogas - 35 - 40%. By purification of the ballast, hydrogen sulfide and drying the heat of combustion of biogas can be increased to 27200 - 31400 kJ / m3 (6500 - 7500 kcal / m3), which is 80% of the calorific value of natural gas. Thus, landfill gas is a renewable resource suitable for use in conventional power plants.

In general, the scheme landfill gas has the form which is shown in Figure 1.

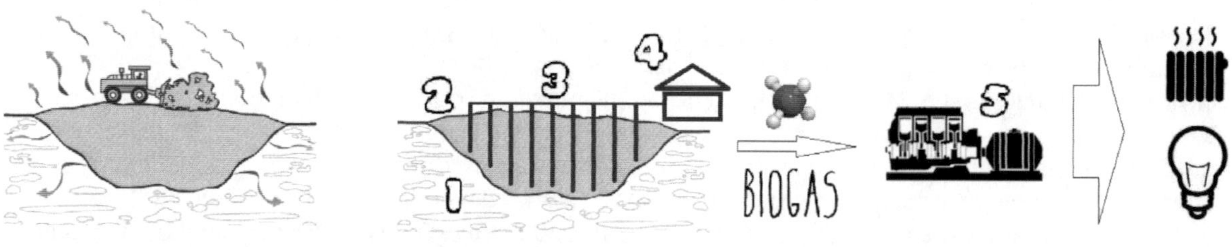

Fig. 1. Typical utilization of landfill gas, where

1 - waste storage area; 2 - landfill body; 3 – degassing system of landfill gas; 4 - section where landfill gas parameters are bringing to the required value; 5 - cogeneration facility

In most cases, systemic study of the morphological structure and the decomposition process of organic compounds has not been since the founding of polygons. In the absence of objective information on the current stage of the life cycle of organic compounds in the body of the landfill to give a qualitative assessment of the energy intensity of the landfill and the economic feasibility of the construction of the energy complex is not possible.

In analyzing the issue of construction of the energy complex we have formulated priorities which allow an objective assessment of the capacity of the landfill in the generation of the landfill gas: stratified analysis of the morphological structure of the landfill on the marker layers (time-projection installation waste); analysis of the current status of the life cycle of organic compounds in the body of the landfill; experimental determination and subsequent monitoring of the environmental parameters (temperature, moisture, pH of the environment, etc.) in the body of the polygon marker layers (the time of the projection 3, 5, 10, 15 or more years of laying waste); experimental determination of the debit layers of the landfill and mine as a whole.

By solving these problems the targeted problem-oriented method of management of solid waste landfill and development of the effective energy complex on it will be developed.

References

[1] Brown K A, Maunder D H 1994 *Using landfill gas: A UK perspective Renewable Energy* 5 (5-8) pp 774-781

[2] Machado S L, Carvalho M F, Gourc J-P, Vilar O M do Nascimento JCF 2009 Methane generation in tropical landfills: Simplified methods and field results *Waste Management* 29 (1) pp 153-161

[3] The global methane emissions and the possibility of reducing GlobalMethaneInitiative (GMI) URL: https://www.globalmethane.org/documents/analysis_fs_rus.pdf (date treatment 19.03.2013g.)

[4] Thompson S, Sawyer J, Bonam R, Valdivia JE 2009 Building a better methane generation model: Validating models with methane recovery rates from 35 Canadian landfills *Waste Management* 29 (7) pp 2085-91

[5] Themelis N J , Ulloa P A 2007 Methane generation in landfills *Renewable Energy* Volume 32 Issue 7 pp 1243-57

[6] I R Gil'manshin 2014 Bio-organic waste Tatarstan: features and prospects of recovery, taking into account the regional dimension *Modern problems of globalization of the world economy and socio-cultural development: Proceedings of the final scientific conference* (Kazan: Fatherland) pp 74-76

[7] Azimov Y I, Galeeva A I, Gil'manshin I R, Gil'manshina S I, Ferenets A V 2014 Innovative ways to recycle industrial and consumer waste to produce products with high performance *Innovative engineering technologies, equipment and materials* Materials Science and Technology Conference. - Part 1 (Kazan) pp 180-184

[8] Multriwell® durable gas ecstraction (Multriwell) URL: http://interactivepdf.uniflip.com/2/74963/290615/pub/ (date of treatment: 19.03.2013g.).

[9] Kashapov N F, Gil'manshin I R, Konahina IA 2014 System analysis of the energy complex of engineering enterprise as a basic tool of effective energy management *IOP Conference Series: Materials Science and Engineering* (Bristol-UK) Vol 69

[10] Gil'manshin I R, Kashapov N F 2014 Energy service contracts in regional engineering center for small and medium businesses *IOP Conference Series: Materials Science and Engineering* (Bristol-UK) Vol 69

[11] Asiya Galeeva, Nafisa Mingazova, Iskander Gilmanshin 2014Sustainable Urban Development: Urban Green Spaces and Water Bodies in the City of Kazan, Russia // Mediterranean Journal of Social Sciences MCSER Publishing, Rome-Italy. Vol. 5, No. 24, November. pp 356-360

[12] Valitov Sh M, Gil'manshin I R 2014 Financial control of the implementation of energy efficiency programs of educational institutions of Kazan *Economic Bulletin* (Kazan: CFI) Vol 3 pp 44-51

[13] Gilmanshin I R, Ferenets A V, Azimov Yu I, Galeeva A I, Gilmanshina S I 2015 Innovative technologies of waste recycling with production of high performance products *IOP Conference Series: Materials Science and Engineering* Vol 86 pp12014-16

[14] Gavrilova V A, Kashapov N F and Kashapov R N 2011 *Biomedical Engineering* **45** 198–200
[15] Dautov G, Dautov I, Fayrushin I and Kashapov N 2013 *J. Phys.: Conf. Ser.* **479** 012001

IOP Conf. Series: Materials Science and Engineering **134** (2016) 012008 doi:10.1088/1757-899X/134/1/012008

Investigation of the motion and heat transfer of water droplets in the swirling air flow in weightlessness

D A Gubaidullin[1], V L Fedyaev[1,2], I V Morenko[1], B A Snigerev[1] and E R Galimov[2]

[1]Institute of Mechanics and Engineering, Kazan Science Center, Russian Academy of Sciences, Kazan, Russian Federation
[2]Kazan National Research Technical University named of A. N. Tupolev-KAI, Kazan, Russian Federation

E-mail: morenko@imm.knc.ru

Abstract. The motion and heat transfer of water droplets with a swirling air flow is investigated. Flow was considered in a cylindrical chamber in the absence of gravity. We created a mathematical model of this problem and made appropriate calculations. The features of the air flow at a tangential feeding it into the chamber, and the motion of the drops, their thermal behaviour are founded. We presented the recommendations for the rational choice of parameters of the apparatus and rational operation regime.

1. Introduction

As a result of the contact of air and other gaseous media with a liquid, this is usually heavily dispersed, located in the gas solid particles deposited on the interface. Combined cycle gas contaminants absorbed by liquid. In the presence of the temperature difference between these mediums the heat transfer takes place. Also, evaporation or condensation of liquid on its interfacial surface are possible.

In one form or another, these processes are implemented in wet scrubbers [1], to some extent, in the cooling towers [2-4]. In the absence of gravity cleaning processes, heat and mass transfer can be implemented in the working chamber of apparatus. In this chamber, tangentially supplied air with impurities irrigated with water. It should be noted that the swirling vortex flow are widely used in engineering: vortex separators, cyclones, precipitators, dryers, cyclic furnaces, burners, plasma devices [5-7].

2. Results of numerical simulations and analytical estimates

Issues of mathematical modeling of isothermal and non-isothermal swirling flows devoted to work [5], [8-9], numerous articles. In this paper, modeling of non-isothermal flows of two-phase medium is carried out using a complete system of equations of turbulent motion, including a standard two-parameter $k-\varepsilon$ turbulence model, initial and boundary conditions. In describing the motion of water droplets applied Lagrange continuum approach. Integration of the corresponding system of equations is carried out numerically by finite volume method.

IOP Conf. Series: Materials Science and Engineering **134** (2016) 012008 doi:10.1088/1757-899X/134/1/012008

The computational domain consists of a cylindrical chamber with a pipe. Total height of the chamber 0.9 m, the diameter of the inlet pipe is 0.07 m (figure 1). The air velocity in the inlet pipe is taken 25 m/s, the temperature 303 K. Water-spraying device is located in the chamber at a distance 0.135 m from the lid. The drops are injected against the flow of air at a rate 1.0 m/s, water temperature 283 K.

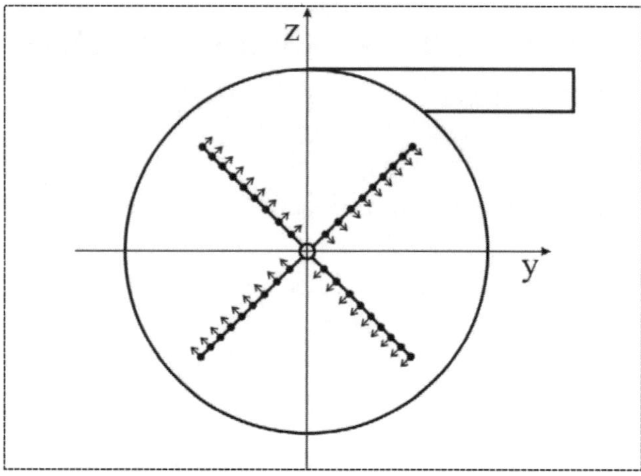

Figure 1. Scheme of the cylindrical chamber. View from above.

We found that due to the tangential supply of air into the chamber it acquires rotary motion. The maximum rate is achieved in the region where the air from the nozzle accumulates on the concave wall of the chamber (figure 2). Further, the stream of air flows over the surface and twists. The minimum value of the circumferential velocity component v_z takes the central part of the chamber, and the maximum near the concave surface.

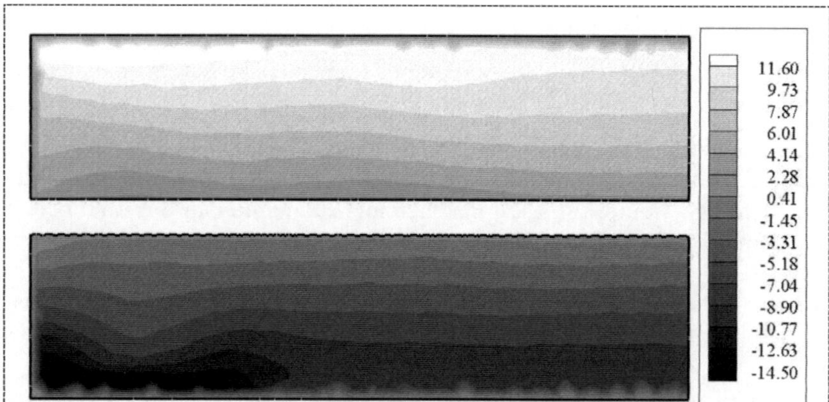

Figure 2. The velocity component v_z in section $z = 0$.

Changes in the velocity magnitude in the cross sections of the chamber $x = 0.2$ m, 0.55 m, 0.85 m illustrated on figure 3. It is evident that the distance from the entrance to the change velocity magnitude in the radial direction becomes more uniform.

We shown that in the central portion of the chamber drops fairly long time moving against the air flow (figure 4). While the droplets, which was ejected on the periphery, almost immediately addicted to the carrier medium. Also, they are changes its original direction of movement. Some of these droplets nailed to the walls of the chamber. From the results of calculations that the average air temperature at the exit of the chamber is about 298 K, wherein the periphery is slightly lower than in the central portion of the chamber.

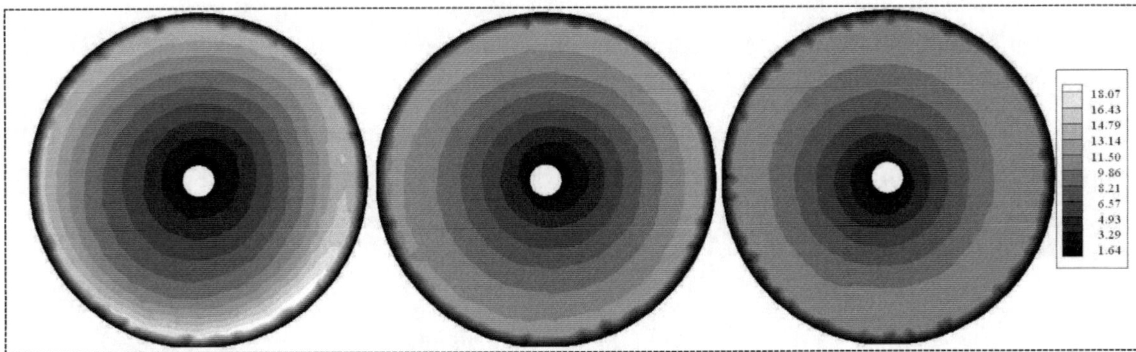

Figure 3. Magnitude of the air velocities in sections $x = 0.2$ m, $x = 0.55$ m, $x = 0.85$ m.

Figure 4. The position of the particles in the stream.

We made further simplifications in order to obtain engineering dependencies for evaluating the main indicators of this apparatus. At the front of the chamber the flow of air and the movement of water droplets is not significantly regular, so they are not considered. We suppose that the working environment (air) - ideal fluid. At the rotational-translational motion in the central part the vortex core is formed. The flow considered a potential outside core on the periphery [5]. However in contrast to [5], where the district v_φ and longitudinal v_z component of the velocity vector of the air have angular points, approximation suggested some excluded them:

$$v_\varphi = \begin{cases} \omega r\left(3 - \dfrac{2r}{r_1}\right), \\ \omega r_1^2 / r \end{cases} \quad v_z = \begin{cases} v_0\left(1 - \left(\dfrac{r - r_1}{r_1}\right)^2\right), & a_0 \le r \le r_1; \\ v_0 & r_1 \le r \le a. \end{cases}$$

Where r – the coordinate in the transverse direction of the cylindrical coordinate system (z,r,φ), r_1 – the radius of the surface section of the vortex and potential flows of air,

$$r_1 = 0.5a\sqrt{3\left(p_\infty - p_0 - 0.5\rho_1 v_0^2\right)/\left(\rho_1\omega^2 a^2\right)};$$

a_0, a – the radius of the central tube and the cylindrical wall of the chamber; ω – the angular velocity of rotation of the air, v_0 – longitudinal velocity on the periphery of the flow associated with the average estimated velocity w_1 by ratio

$$v_0 = w_1 a^2/\left(a^2 + r_0^2 - r_1^2/6\right);$$

ρ_1 – the density of air; p_∞, p_0 – the pressure at infinity and the center of the vortex (at the central part of the chamber). According to the law of conservation of momentum ω determined from the equation

$$\omega = u_1 G_1\Big/\left[2\pi w_1 r_1^2\left(a - \frac{\omega r_1^2}{\chi_1}\right)\right],$$

where u_1 – the average speed in the inlet pipe of the working chamber, G_1 – the air flow rate,

$$\chi_1 = \sqrt{3\left(p_\infty - p_0 - 0.5\rho_1 w_1^2\right)/\rho_1}.$$

As a result of integrating of the drop motion equations wherein the drag force is determined by the Stokes formula, we found that the longitudinal speed u_p close to the air velocity v_z. The drop velocity v_p in the transverse direction along the axis r, described by the relation:

$$v_p = \left[v_{p0} + (1+\chi)^2 \overline{v}_\varphi^{2}\right]\Big/\left[R_\mu\left(r_0^2 + 2\overline{v}_\varphi^{2}\left(1+\chi^2\right)\tau/R_\mu\right)^{1/2}\right].$$

Where τ – the time, r_0, v_{p0} – the initial coordinates, the drop velocity; $\overline{v}_\varphi^{2}$ – averaged over the cross section of the circumferential speed of the air, $\chi = \left(w_{p0} - \overline{v}_\varphi\right)/\overline{v}_\varphi$, w_{p0} – the initial circumferential velocity of the drop, $R_\mu = 6\pi r_p \mu_1/m_p$; r_p, m_p – the radius, mass of the drop. The drop peripheral velocity $w_p = \left(w_{p0} - v_\varphi\right)\exp\left(-R_\mu\tau\right) + v_\varphi$.

Considering further drop of thermally thin body, its temperature determine

$$\theta_p = \tilde{t}_1 + \left(\theta_{p0} - \tilde{t}_1\right)\exp\left(-\tau K_\alpha\right).$$

Where θ_{p0} – the initial volume average temperature of the drop, $\tilde{t}_1$ – the average temperature in the considered part of the chamber,

$$\tilde{t}_1 = \frac{2c_1 G_{1m}\tilde{t}_{11} + c_2 G_{2m}\theta_{p0}\left(1 - \exp\left(-\tau K_\alpha\right)\right)}{2c_1 G_{1m} + c_2 G_{2m}\left(1 - \exp\left(-\tau K_\alpha\right)\right)},$$

$\tilde{t}_{11}$ – the average air temperature at the inlet of the selected portion of the chamber; G_{1m}, G_{2m} – the mass flow of the air, water in the chamber respectively; c_1, c_2 – the specific heat capacity of air, water; $K_\alpha = \alpha_p s_p/\left(m_p c_2\right)$, s_p – the drop surface area.

At the initial air temperature $\tilde{t}_{11}$=303 K, water θ_{p0}=283 K, the flow rate G_{1m}=0.1 kg/sec, G_{2m}=0.04 kg/sec, $c_1 = 1.0\cdot10^3$ J/(kg·°C), $c_2 = 4.2\cdot10^3$ J/(kg·°C) and the flight time of the drop τ_{II}=1.5 sec find $\tilde{t}_1$=301 K. Consequently, the output from the chamber average temperature $\tilde{t}_{12}$=299 K, the water temperature is about 285 K.

3. Conclusions

Thus, on the basis of the results we made the following conclusions.

1. Initially, there is substantially unequal air flow. Drops motion away from the inlet stabilized fairly quickly. This fact is the basis to apply the analytical methods.
2. To reduce the proportion of the drops deposited on the walls, better to make the chamber as a conical shape. Furthermore, it can be achieved by increasing the pipe diameter.
3. Generation of acoustic waves in the chamber provides an intensification of heat and mass transfer. Also, it will allow to reduce the temperature of the air.

Acknowledgments
This work was supported by the Program OEMMPU Russian Academy of Sciences "Fundamental problems of physical and chemical mechanics for experiments on the International Space Station".

References

[1] Aliev G M 1986 *Technology dedusting and purification of industrial gases* (Moscow: Metallurgy)
[2] Ponomarenko V C and Arefiev J I 1998 *Cooling towers of industrial and energy companies* (Moscow: Energoatomizdat)
[3] Laptev A G and Vedgaeva I A 2004 *Design and calculation of industrial cooling towers* (Kazan: KGEY)
[4] Fedyaev V L, Mazo A B, Morenko I V, Gainullin R F and Gainullina R F 2009 *Problems of energy* **1–2** 15
[5] Goldshtik M A 1981 *Vortex Flows* (Novosibirsk: Nauka)
[6] Galyautdinov R T, Kashapov N F and Luchkin G S 2002 *Inzhenerno-Fizicheskii Zhurnal* **75** 170
[7] Saifutdinov A I, Fadeev S A, Saifutdinova A A and Kashapov N F 2015 *JETP Lett.* **102** 637
[8] Mitrofanova O V 2010 *Hydrodynamics and heat swirled flows in channels of nuclear power plants* (Moscow: Fizmatlit)
[9] Tarasevich S E, Fedyaev V L, Yakovlev A B and Morenko I V 2012 *Int. Conf. on Nanochannels, Microchannels and Minichannels* **HT 2** 109

Features of thermal processes of plasma deposition and hardening of coatings with external modulation parameters of the arc

A M Kadyrmetov[1], **S N Sharifullin**[2], **A F Maltsev**[1]
[1] Voronezh State Forestry academy, 394613, st. Timiryazev, house 8, Voronezh, Russia
[2] Kazan Federal University, 42000818, Kremlyovskaya str., house 18, Kazan, Russia

E-mail: Saidchist@mail.ru

Abstract. In the work on the basis of mathematical modeling analysis of processes of plasma deposition of coatings with modulation of the electrical parameters of the extension arc. The effect of modulation on the temperature field in the system "coating-basis" on a local scale, proportionate to the diameter of the spot attachment of the arc to the surface, and at the macro-level of evaporation surface. It justifies the preconditions of the improvement of plasma deposition and hardening coatings.

1. Introduction

Current status of plasma technologies of coating and surface hardening shows that the traditional ways of their development on the basis of modifications of stationary processes have reached the limits of a marked improvement in their effectiveness. Analysis of ways to improve plasma deposition and hardening coatings shows that effective improvement is possible by using the methods of dynamic modification of parameters and, in particular, by modulating the electrical parameters of the external arc (straight arc) [1, 2].

The process of plasma spraying and hardening coatings (Pnip) through the use of dynamic effects in the modulation of the electrical parameters of the plasma torch is one of the promising and, however, poorly understood directions. The object of the theoretical analysis Pnip when the modulation of the electrical parameters of the plasma torch remote arc was thermal and physico-mechanical processes in the system "plasma jet and remote arc – coating – basis". The task was solved by development of mathematical models [3-5] and subsequent simulation of the above objects of study in the modulation of the electrical parameters of the remote arc plasma torch.

The effect of modulation of electrical parameters on temperature field evaporation surface has two space-time dimensions. In aspect local impact on the coating to improve the bonding strength of the coating to the substrate is necessary to provide a discretely distributed local zones of melting of the surface to the transition zone of the "coating-substrate". This is true in particular for coating under dynamic loads [6, 7]. In the aspect of scale of all surfaces is necessary to minimize variation of temperatures on the profile of the substrate surface and the uniformity of its temperature field.

To develop process recommendations in the processes of plasma deposition and hardening coatings requires knowledge of regularities and links between technological factors in the production of coatings in the mode of modulation of external parameters of the arc and thermal criteria of quality of coatings, determining the level of residual stresses in coatings and their mechanical and tribological properties.

To perform this task have been developed mathematical models of thermal processes describing the processing system of the process of applying and hardening coatings in the modulation mode of the external parameters of the arc.

2. Methods of Research

Methodological basis of theoretical research formed the scientific basis of the theory of thermal welding processes, methods of computational mathematics (finite difference method), modern graphics and computing systems to computers.

3. Mathematical model of thermal processes in the system "coating-basis"

Physical processes in the coating during modulation of the power transferred arc plasma torch with ΔN_p amplitude and time of impulse τ_i consist in fusing the coating to the transition zone to the substrate, and the formation of physical contact at time t_c, and the chemical interaction of the coating with the base point in the local zone of direct binding of the arc to the base (coating). Across the surface taking into account the influence of speed V and feed S of the plasma torch formed certain distribution density of the discrete zones of melting and of thermal effect on the base and the coating, causing the uniformity of the temperature field base and to increase the physico-mechanical and tribotechnical characteristics of coatings. To estimate the thermal efficiency of the modulation of the electrical parameters during plasma spraying of coatings was constructed mathematical model of the temperature field T of the system "coating-substrate", which is represented as the sum of the total field away from the source T_Σ, local field near T_r and the initial temperature T_0 [2, 8]:

$$T = T_\Sigma + T_r + T_0 . \qquad (1)$$

To solve the problem, following assumptions were:

1. The power of the heat source is constant during the pulse (N_{imp} = const, q_{imp} = const).

2. The heat source is concentrated in the elementary point; the error estimation of the temperature from this to a depth equal to the thickness of the coating, can be considered.

3. The heat source is uniformly and rectilinearly moves product at a constant speed (V = const).

4. The distribution of heat in the body occurs according to the law of heat conduction Fourier.

5. The boundary of the body is impermeable to heat transfer from the body.

6. The coefficients of the thermophysical properties of the base metal (coefficient of thermal conductivity and volumetric heat capacity) do not depend on temperature.

7. Phase and structural transformations of the metal coating occur without the allocation and absorption of heat.

8. Initial and boundary conditions: $q\big|_{\tau\geq 0} = q$; $\left.\dfrac{\partial T}{\partial z}\right|_{z=0} = 0$, $T\big|_{z=\infty} = 0$.

Then the total field away from the source T_Σ will be installed Rikaline N. N. [9], and describe the limiting state of heat conduction in semi-infinite body heat without moving relative to the heat source on the surface in the direction of the $0X$ axis with constant speed V:

$$T_\Sigma(R, x, \infty) = \frac{q}{2\pi \cdot \lambda \cdot R} \cdot e^{-\frac{V \cdot x}{2a} - \frac{V \cdot R}{2a}}, \qquad (2)$$

where R is the radius-vector from the heat source, m; ; x, y – coordinates of the surface of the body, m; z is the coordinate directed into the body perpendicular to its surface, m; V – velocity of the heat source, m/s; a – thermal diffusivity, m^2/sec; λ– thermal conductivity, W/(m·deg).

When $x = 0$ and $y = 0$ we have $R = z$ is the depth of the foundations of the body.

To determine the heat flux to the coating, necessary for its local penetration, and maximum temperature T_{max}, equal to or greater than the melting temperature of the coating material, the expression (1) has the form:

$$q_{imp} = \frac{4{,}25 \cdot V \cdot \rho_m \cdot c_m \cdot z^2 \cdot T_{max}}{\xi \cdot \varphi(\rho, \tau)}, \tag{3}$$

where V – speed of movement of plasma generator, m/s; ρ_m – the density of coating material, kg/m3; c_m is specific heat of the coating material, j/(kg•°C); ε – factor distribution of the heat source:

$$\varepsilon = 2\frac{h}{r_0} \cdot \left[\sqrt{(h/r_0)^2 + 1} - \frac{h}{r_0}\right]. \tag{4}$$

The total power transferred arc N_h (W) pulse with modulation, ensuring the penetration of the coating at a depth h (m) to the transition zone to the substrate, was estimated using the expression

$$N_h = 4{,}25 \cdot V \cdot c_V \cdot h^2 \cdot T_{пл} / \varepsilon, \tag{5}$$

where V – speed of movement of the spot binding the straight arc surface, m/s; cv – volumetric heat capacity of the coating material, j/(kg•K); T_p– melting point material coating, °C; r_0 is the radius of the spot of the direct binding of arc to the coverage, m.

For cylindrical bodies, sprayed on the helix, the common box away from the heat source TΣ expression was determined on the basis of theoretical provisions [10]:

$$T_\Sigma(r, z, \theta, t) = \frac{q_n}{R^2 \cdot C_\gamma} \cdot \frac{1}{\sqrt{\pi}} \cdot \sum_{n=1}^{N} \frac{\Phi(\rho, \tau_n)}{\sqrt{\tau_n}} \cdot \exp\left[-\frac{\xi^2 \cdot n^2}{4 \cdot \tau_n}\right], \tag{6}$$

where $\tau_n = \frac{2\pi \cdot a}{R \cdot V_\theta} \cdot n$; $\xi = \frac{H}{R}$; $q_n = \frac{q}{V_z}$; τ_n is the dimensionless complex which characterizes the time; ξ is the dimensionless complex, which characterizes the spraying step; θ is the angular coordinate, rad.; q – thermal power of the plasma torch, W; q_n – linear energy, j/m; R – radius of cylinder, m; Φ(ρ, τп) – the function characterizing the process of propagation of heat along the radius of the cylinder; n – number of turns; N = 1, 2, 3, etc.; V_z, V_θ, respectively, the axial and circumferential speed of the heat source, m/s; C_γ – volumetric heat capacity, j/(m^3•grad); H – pitch of the, m.

The maximum temperature T_m was determined from the dependence:

$$T_m = \frac{2 \cdot \varepsilon \cdot q}{\pi \cdot e \cdot V \cdot C_\gamma \cdot r_x^2}, \tag{7}$$

where r_x is the local radius-vector pointing from spot spraying the body parts with coating, m; ε– factor distribution of the heat source.

The density of the thermal потокаq in the case of modulation of the parameters of the plasma torch is equal to

$$q = q_0 + q_{imp}, \tag{8}$$

where q_0 – the load capacity of indirect and remote arc; q_{imp}– weighted average effective thermal power of the arc remote from the modulating pulses.

Weighted average effective heat output of the modulating pulses is determined by the formula:

$$q_{imp} = \Delta q_{imp} \cdot \tau_{imp} \cdot \nu_{imp}, \tag{9}$$

where Δq_{imp} – amplitude of power pulses; τ_{imp}– pulse duration; ν_{imp}– pulse repetition rate.

To solve the heat problem on the estimation of the pulsed power arc N_{imp} remote from the penetration depth of $h + \Delta h$ and the speed of the torch V, and the dependence of the number of local penetration zones of the n_p, the processing step H, the pulse width t_{imp} and the modulation frequency v_m from the above specified parameters were made given the following options:

– coating thickness h in the range of 0.1-1 mm and Δh of the transition zone in the range (0,05 ... 0,1)h;

– specific heat capacity cm and density ρ of the material of the coating;

– the size of the local penetration of the transition zone of the coating to the Sml basis;

– the area of the base surface S_b;

– the speed of motion *V* of the plasma torch;
– duty cycle *k* $(k > 1)$;
– the proportional limit of the material covering σlp;
– operational stress σ_e.

4. Results and discussion

On the basis of decisions (2) for semi-infinite bodies and (5) to the cylindrical body using a method of reflection and superposition (superposition) of the thermal waves, but also taking into account processes of heat and equalization of the medium-integral increase of body temperature were obtained for solutions of the relevant bodies, such as, for example, the tooth of the excavator and the crankshaft. A mathematical model of the temperature field of the system "coating-substrate" is implemented numerically.

In the aspect of the scale of the entire surface of the control parameters external modulation allows the arc to minimize variation of temperatures on the profile of the substrate surface and the uniformity of its temperature field. This is confirmed by the results of calculations for cylindrical parts when plated on the helix (Fig. 1). From the graph of dependence of temperature basis T_Σ from the distribution of the average amplitude of the power q_{imp} and the rotational speed *n* can be seen that the temperature spread in the heating output of the pulses is from 2 to 5 % of the average. The traditional use of preliminary heating with the plasma torch without longitudinal feed along the cylindrical temperature spread is 10 to 20 percent of the average wage or more, which is 80 – 180 °C more than by using external modulation of the arc. The decrease of the temperature spread on the surface of the part reduces the difference of temperature deformations, creates conditions for a uniform distribution of structure and physico-mechanical properties of coatings and basis. The results of calculations for example the crankshaft of the engine KAMAZ-740 showed that the difference in surface temperature of the sputtering is not more than 2...5 °C is provided by adjusting the weighted average effective heat capacity q_{imp} by its reduction on the first three threads in the range from 165 to 10 watts for the connecting rod journals and in the range from 370 to 160 watts for indigenous journals.

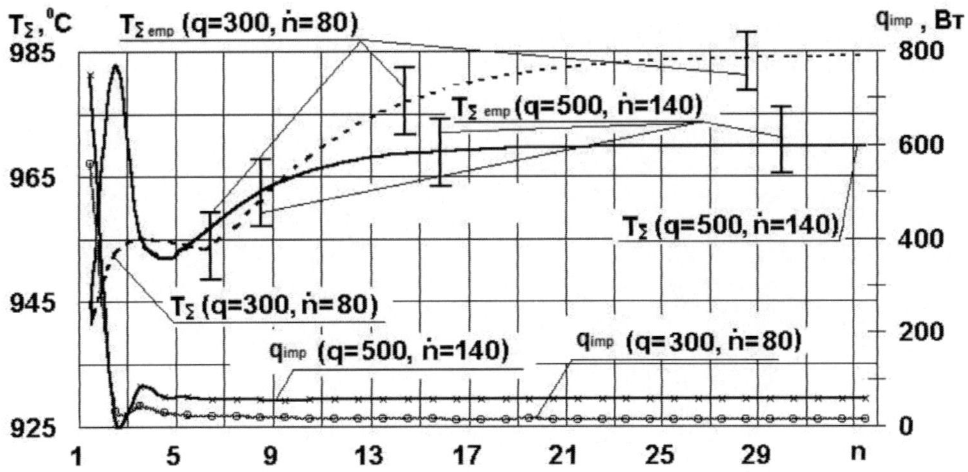

Figure 1. The dependence of the temperature T_Σ basis of the frequency of rotation *n* and the distribution of the weighted average effective heat capacity q_{imp} if modulation power extension arc: R = 0.04 m; L = 0,002 m.

The modulation power extension arc by sputtering on the inner surface of the small-size also helps to ensure the uniformity of the temperature field evaporation surface. Even more relevant to use the power modulation of the extension of the arc during the deposition profile on the surface of parts subjected to high wear stresses, such as working bodies tillage, forestry and road construction machinery. In particular, it refers to the wedge-shaped parts such as bucket tooth of excavator, plough, cultivator. On their surfaces the uneven heating may reach 600 ° C. The modulation power extension arc allows to reduce this value to 10 to 20 °C.

The results of the calculation of the heat flow to the nichrome coating (H80X20), similar in composition to the material self-fluxing Nickel-base PG-CP4, which provides the penetration of the coating to the depth of the transition zone with the base presented in Fig. 2, 3 [11]. Analysis of the curves shows that with increasing thickness of the coating is 10 times (from 10^{-4} to 10^{-3} m), the required heat flux to the surface must be increased by 1 – 2 orders of magnitude (in the velocity V = 0,2...0,8 m/s). Increase the speed of the plasma torch from 0.2 to 0.8 m/s increases the required heat flux to the surface of 1.5 – 2.5.

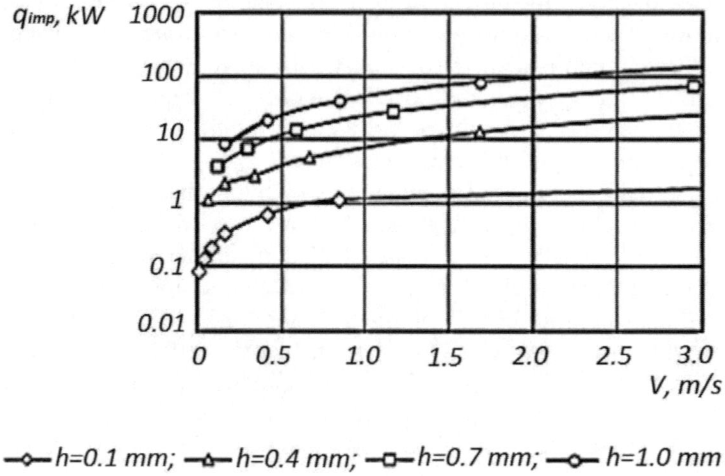

Figure 2. The dependence of the heat flux to the surface q_{imp} when the pulse power remote arc, ensuring the penetration of the coating to the transition zone with the base, on the speed of the torch *V*

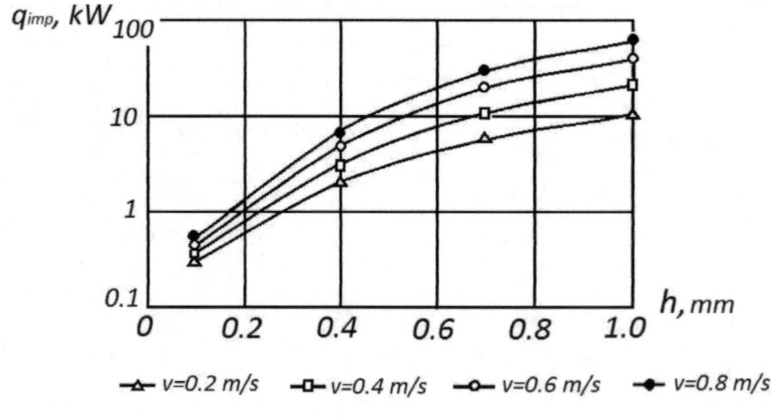

Figure 3. The dependence of the heat flux to the surface q_{imp} when the pulse power remote arc, ensuring the penetration of the coating to the transition zone to the substrate, the thickness of the coating *h*

It is known that the formation of a strong connection of the melt with the solid phase is only possible as a result of physical contact and chemical interaction between them. This can be achieved if

IOP Conf. Series: Materials Science and Engineering 134 (2016) 012009 doi:10.1088/1757-899X/134/1/012009

the contact time between the phases is not less than the sum of the delay time of diffusion transitions and development processes heterodiffusion before the formation of supersaturated solid solutions. It is known that the diffusion processes during solidification of metals are three stages with characteristic time constants of $10^{-12} - 10^{-10}$, $10^{-8} - 10^{-5}$ and $(0,5 - 1,5)10^{2}$ s, corresponding to oscillation periods of atoms and translational switch groups of atoms, the oscillation periods of the clusters and displacement of layers of clusters. This suggests that, at least temporarily condition the formation of a firm connection is ensured when the pulse over 10 ms. If enough are the first two stages of crystallization, this time should be more than 10 iss, which requires confirmation in additional studies.

Consider the results of the calculation for material H80X20 and the selected source of data: c_m = 750 j/(kg• K); ρ_m = 8400 kg/m3; $a = 1,65{\bullet}10^{-5}$ m²/s; σ_e = 300 MPa; σ_{Ip} = 6.5 GPa; $T_{max} = T_{mel}$ = 1400 °C; h = 0.1-1 mm; Δh = (0,03 0,05...)h; V = 0,001-0,8 m/s; S_b = 0,0168 m²; $S_{ml} = 1,3{\bullet}(10^{-8}...10^{-6})$ m²; k = 4 – 6. Heat flux to the surface, ensuring penetration of the coating to the transition zone to the substrate, when such data can be evaluated according to the graphs in Fig. 2, 3.

The use of power modulation of the plasma torch remote arc should ensure the penetration of the coating to the depth of the transition zone with the base $h + \Delta h$ at evenly distributed plots, each S_f during the movement of the plasma torch at a speed V (Fig. 4, 5). The cross section of each plot is close to a semicircle. The total number of zones with a penetration coating must be selected on the basis of the factor of safety that exceeds the operating load surface.

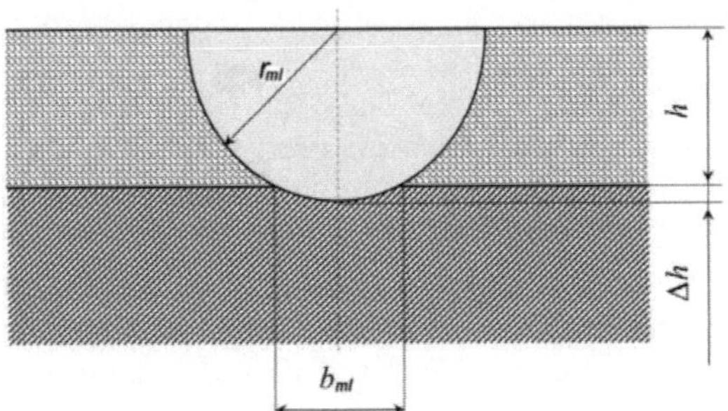

Figure 4. The transverse cross section of the molten zone of the coating formed from the pulse power transferred arc: r_{ml} – the radius of the molten zone surface; h is the coating thickness; Δh is the thickness of the transition zone; b_{ml} – width of the molten transition zone

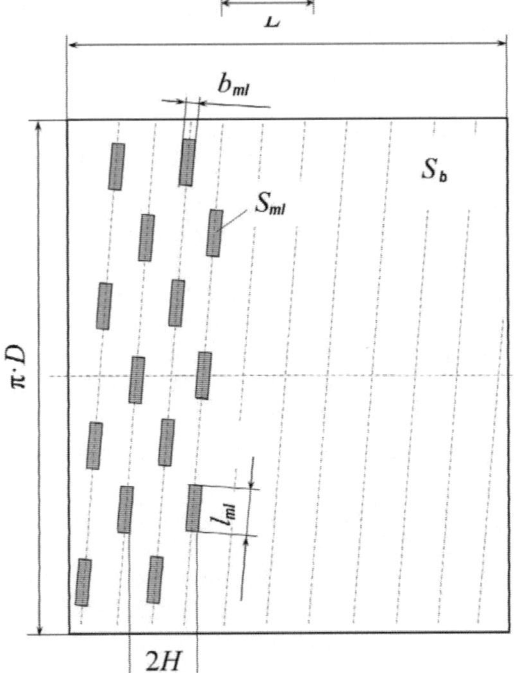

You want to find the dependence of the pulsed power arc N_{imp} remote from the penetration depth of $h + \Delta h$ and the speed of the torch V, and the dependence of the number of local penetration zones is N, the processing step H, the pulse width t_{imp} and the modulation frequency ν_m from the above specified parameters.

Figure 5. Scan a cylindrical surface and arrangement of the fused areas of the coating: D, L – diameter and length of cylinder, respectively; Sb is the area of the cylindrical surface; S_{ml}, l_{ml}, b_{ml} – area, length and width of the transition zone of the molten coating to the substrate, respectively; H is the step helix machining

The number of local zones of melting is in the range from 3432390 pieces for the coating thickness is 0.0001 m (S_{ml} = 1,34•10 8 m²; $b^{ml} = 6,4{\bullet}10^{5}$ m; lml = to 0.00021 m; H = 0.35 mm) to 34324 pieces for a coating thickness of 0.001 m (S_{ml} = 1,34•10 6 m²;

b_{ml} = 0,00064 m; l_{ml} = 0,0021 m; H = 3.5 mm). The pulse duration is the interval from 75 iss to 37,5 ms (modulation frequency of v_m = 20...10000 Hz).

Experimental studies have shown that modulation of the extension of the arc straight polarity allows to obtain coatings with high physical-mechanical and tribotechnical properties. In comparison with the coating without direct modulation of the arc: the bond strength of the coating to the substrate is increased to 1,15...1,25 times, microhardness is 1.1...1.2 times, fatigue resistance of the samples to 1.2 times the wear resistance of coatings – 1.25...1.35 times. The porosity of coatings decreases 1.2...1.3 times. Improving the basic physical and mechanical properties of the coatings allows the use of the developed technology for components operating at alternating and cyclic loading.

4. Conclusions

The mathematical model of thermal processes in plasma spraying of coatings with pulse modulation external arc plasmatron allows to calculate the parameters of power amplitude and pulse width and frequency modulation, providing local penetration of the coating to the transition zone to the substrate, and a uniform temperature field. The dependence of the power of the arc remote from the thickness of the coating identical to that for the case of stationary heat source. Possible calculation error can be eliminated by conducting experiments clarifying.

The results of calculations allow to determine the optimal spacing of the heat action, which provides higher strength characteristics of coatings in comparison with conventional plasma spraying.

References

[1] Kadyrmetov AM. The Management of technological providing of processes of plasma deposition of coatings in the mode of electrical parameters modulation: monograph /A M Kadyrmetov. – Voronezh: Scientific book, 2013. – 260 p.

[2] Stanchev DI. Methods modeling of thermal processes during the deposition and fusing of parts of complex shape /DI Stanchev, AM Kudermetov, VA Vinokurov, VN Bukhtoyarov //Mathematical modeling, computer optimization technology, equipment and management systems the forest sector: Interuniversity collection of scientific papers. – Voronezh: Voronezh state forestry engineering ACADEMY, 2001. – Pp. 161-164.

[3] Saifutdinov A I, Saifutdinova A A, Kashapov N F and Fadeev S A 2016 *J. Phys.: Conf. Ser.* **669** 012045

[4] Dautov G, Fayrushin I and Kashapov N 2014 *J. Phys.: Conf. Series* **567** 012006

[5] Saifutdinov A I, Fadeev S A, Saifutdinova A A and Kashapov N F 2015 *JETP Lett.* **102** 637–642

[6] Zaripov R G, Kashapov N F, Tkachenko L A, Shaydullin L R 2016 *J. Phys.: Conf. Series* **669** 012053

[7] Kornienko E E, Lapushkina E J, Kuzmin V I, Vaschenko S P, Gulyaev I P, Kartaev E V, Sergachev D S, Kashapov N, Sharifullin S, Fayrushin I 2014 *J. Phys.: Conf. Series* **567** 012010

[8] Kadyrmetov AM. Estimation of temperature wedge-shaped surfaces in the process of plasma spraying /AM Kadyrmetov //Technology software engineering industries: Interuniversity collection of scientific papers. – Vol. 2. – Voronezh: VSTU, 2002. – Рз. 55-59.

[9] Rykalin NN. Thermal fundamentals of welding /NN Rykalin. – M. – L.: Publishing House Of The Academy Of Sciences SSSR, 1947. – 272 p.

[10] Makhnenko VP. To the calculation of the temperature field in arc welding of circular cylinder /VP Makhnenko //Automatic welding. 1961. – No. 12. – P. 34 39.

[11] Kadyrmetov AM. Hardening of plasma sprayed coatings by pulse modulation of the plasma torch remote arc /AM Kadyrmetov, EV Smolentsev, AF Maltsev, GA Sukochev //Bulletin of Voronezh state technical University. – 2014. – Vol. 10. – No. 1. – P. 336-341.

Efficiency of surface cleaning by a glow discharge for plasma spraying coating

A M Kadyrmetov[1], N F Kashapov[2], S N Sharifullin[2], A I Saifutdinov[2] and S A Fadeev[2]
[1] Voronezh State Academy of Forestry and Technologies
[2] Kazan Federal University

E-mail: kadyrmetov.a@mail.ru

Abstract. The article presents the results of experimental studies of the quality of cleaning steel surfaces by a glow discharge for plasma spraying. Shows the results of measurements of the angle of surface wetting and bond strength of the plasma coating to the surface treated. The dependence of the influence of the glow discharge power, chamber pressure, distance between the electrodes and the processing time of the surface on cleaning efficiency. Optimal fields of factors is found. It is shown increase joint strength coating and base by 30-80% as a result of cleaning the substrate surface by a glow discharge plasma spraying.

Key words. Glow discharge, cleaning, activation, activation, surface, substrate, experiment, plasma spraying, plasma torch, plasma coating, optimization.

1. Introduction

Preparation of the substrate surface prior to application of plasma coating is one of the most important groups of factors that influence the formation of the coating [1]. The presence of the oxide film on the surface of steel reduces its surface energy and thus prevents the formation of the most complete atomic bonds with the applied coating [2]. For plasma coatings deposited plasma generators in an open and controlled atmosphere, it creates one of the most urgent problems of increasing the strength of the connection of plasma coating to the substrate. Prior to application of such coating a surface prepared in various ways, of which the most commonly used are the chemical surface treatment and sandblasting of the surface. These methods have disadvantages. The chemical surface treatment method is associated with environmental problems. The disadvantage of sandblasting associated with the presence of residues of ultrafine dust embedded in the surface layer. Analysis of available data shows the possibility of overcome these deficiencies and significant increase in bond strength coating to the substrate when it is processed glow discharge [3, 4]. However, presently unknown mechanisms of the processes explained such influence on the strength of the coating to the substrate. Therefore urgent is the definition of the main factors and their impact on the effectiveness of the surface cleaning.

The study aims, firstly, to determine the influence of factors of surface treatment by a glow discharge on the effectiveness of cleaning. Second evaluation of the impact of the quality of cleaning on the strength of the plasma coating to the metal substrate.

2. The experimental technique

The quality of treatment was evaluated by the contact angle of wetting the surface. The angle is measured at the sample surface after its cleaning, filling with argon vacuum chamber to normal pressure, and cooling the sample. For measurements distilled water was used, the droplet size was 2-2.5 mm in diameter. The drop was deposited to the surface in stages: the drop squeezed out of the capillary, and then it is brought to the surface to the ground, then the capillary taps. The angle was measured after 10 minutes.

For surface treatment used glow discharge mode of abnormal combustion. This allows a wide range to change the power and, therefore, energy of the bombarding ions at different gas pressures in the chamber. For ion cleaning, etching diode scheme used, the cathode was treated samples of steel 45 with a working surface of 14 mm diameter.

The working medium is argon. Therefore, the main processing mechanism served as a natural spray that is characterized by the coefficient of dispersion. The coefficient of dispersion essentially depends on the pressure in the treatment chamber and at pressures greater than 13.3 Pa decreases. This is due to the formation on the treated surface dense adsorbed layers, delaying and preventing the dispersal of the material of the surface [5]. The lower limit of the pressure is determined by the conditions of existence of a glow discharge at a given diode circuit. For this reason, in this work the gas pressure is set within the range 1...14 Pa.

The main factors of the process is the pressure, the discharge power, the distance between the electrodes and the processing time. Discharge power was fixed at a predetermined level by controlling the current-voltage characteristics of the power supply. Voltage and current values determined by the power and voltage-current characteristics of the discharge. The voltage-current characteristics of the discharge depends on the pressure in the chamber and the distance between electrodes.

In the abnormal discharge mode power value is selected from the conditions of effective surface treatment. The discharge voltage has changed in the range of 500-5000 V, current densities of 0.2-1.5 mA/cm^2 [6]. At voltages less than 500 V significantly reduces the efficiency of cathode sputtering. And at voltages over 5000 micro arcs appear in the electrode gap. These intervals changes of current density and discharge voltage correspond to change of the specific discharge power in the range of 1-5 W/cm^2. The distance between electrodes was varied in the range of from 5 to 15 mm, the processing time – in the range from 30 to 600 seconds.

On the basis of these data and preliminary experiments to find the empirical dependence of the angle of wetting the surface of the above factors was built of three-factor experimental design [7]. Values of the factor levels and their variation ranges are shown in Table 1 and Fig. 1.

Table 1. Factors their levels and ranges of variation.

Name factor	Unit of measure ment	The code mark	Intervals of variation	Levels factors				
				*+	upper (+)	main (0)	lower (-)	*-
Specific power, W	W/cm^2	X_1	1,34	4,604	3,69	2,35	1,01	0,096
The distance between the electrodes, L	mm	X_2	2	11,36	10	8	6	4,636
Chamber pressure, P	Pa	X_3	3	11,55	9,5	6,5	3,5	1,455

As the plan of the experiment was taken a central rotatable composite plan of the second order with the kernel as a complete three-factor plan, six experiments in star points and six in the center of the plan.

The quality of the surface cleaning was determined by the strength criterion compound coating to the substrate. Plasma spraying samples was performed through certain time after treatment. Coating on the part of the samples was carried out immediately after the treatment. In this case, to protect the surface from oxidation unused portable container. The remaining samples were in the air for a while before coating deposition.

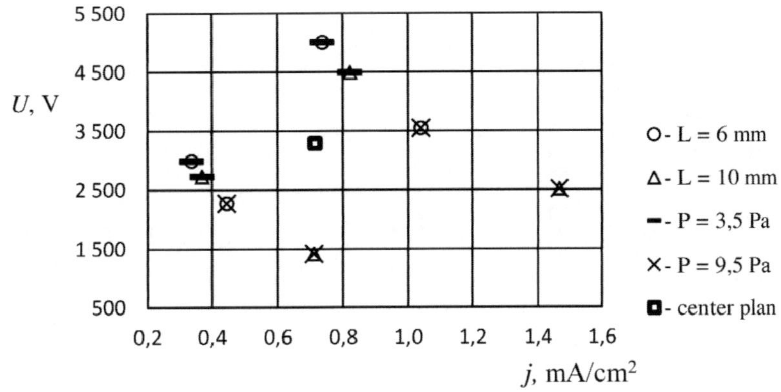

Figure 1. The values of the factors in the core of the plan of the experiment.

Plasma spraying was carried out using an arc plasma torch EDP-104 in the mode: current is 110 A, voltage is 250 V, plasma-forming gas is nitrogen, with the addition of propane, transporting gas is nitrogen, spraying distance is 160 mm. Spraying material is powders self-fluxing material nickel-based PG-HN80SR4, nickel-titanium and nickel-aluminum intermetallic PN55T45 and PN85YU15 dispersion of 40 and 63 microns, respectively. In all cases, deposition time is selected so that the coating thickness was 0.6 ... 0.7 mm, which meets the criterion of correctness [8]: $r / \delta \leq 2$, where r is radius end of the pin; δ is coating thickness. Spraying time was ~ 8-10 seconds.

To quantify the strength of the bond coating to the base method is used stretching taper pin [6] in the centering device on the tensile testing machine MP-05.

3. The results of experimental studies the quality of cleaning steel surface contact angle criterion

Processing the results gave regression equation of the contact angle θ of the factors which is a dimensionless normalized relationship

$$\theta = 22{,}5486 + 1{,}6875X_1 + 1{,}1875X_3 + 3{,}5547X_2X_3 + \\ + 5{,}43634X_1^2 + 1{,}24363X_2^2 + 5{,}07176X_3^2, \quad (1)$$

where $X_1 = \frac{(W-2{,}35)}{1{,}34}$; $X_2 = \frac{(L-8)}{2}$; $X_3 = \frac{(P-6{,}5)}{3}$.

Verification by the Fisher test showed the adequacy of the regression equation (1): $F_p < F_т$ where the estimated value of the Fisher criterion $F_p = 4{,}6$ and the table - $F_т = 5{,}1$ at 5% significance level. Calculated according to the equation (1) the value of θ differ from the experimental to the value not exceeding the experimental error of 10%, which is also confirmed by additional experiments. Therefore, the equation in this region of the factor space can be used as an interpolation formula for calculating the value of the contact angle with an accuracy of 10%.

Optimal values of the contact angle θ, defined by equation (1) is the area of factors: specific power discharge 1,88-2,4 W/cm^2 (~3100-3400 V discharge voltage, the current density ~ 0,60-0,72 mA/cm^2). The residual pressure in the vacuum chamber is 5,5-6,0 Pa; the distance between the electrodes is 8-9 mm. Given the experimental error contact angle at an optimal processing mode was θ = 22±2 degrees.

Wetting angle θ depending on factors surface treatment process constructed according to the regression equation (1), plus additional experimental values of θ in the optimum factors are presented in Figures 2-4. These results show the correspondence between the experimental data and the values calculated by the regression equation (1), with an error of not more than 10%.

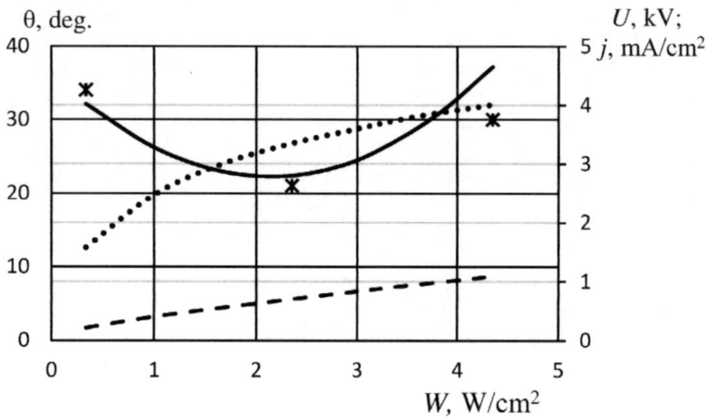

Figure 2. The dependence of the contact angle θ of the specific power W glow discharge: ∗ – experiment (P = 6 Pa, t = 300, L = 9 mm); —— – equation (1); - - - - j; ······ – U.

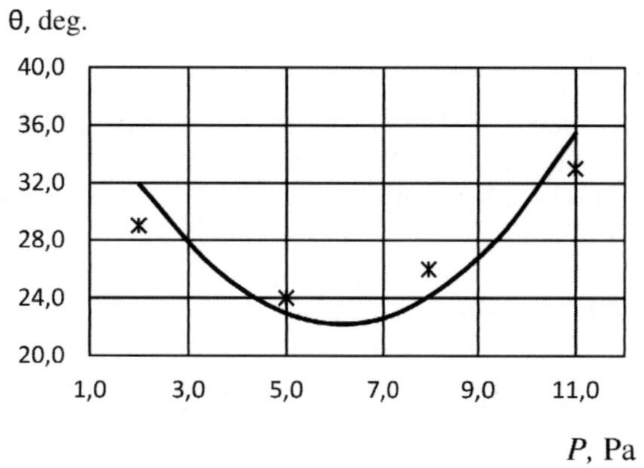

Figure 3. The dependence of the contact angle θ of the pressure P: ∗ – experiment; W = 2,15 W/cm^2 (j = 0,66 mA / cm2; U = 3300); L = 9 mm, t = 300 s.

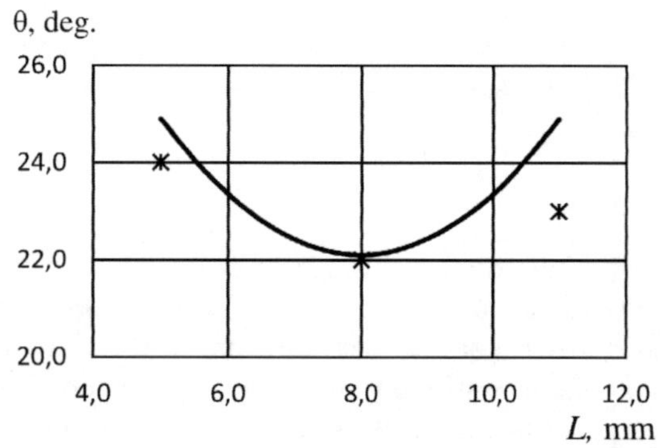

Figure 4. The dependence of the contact angle θ of the distance between the electrodes L: ∗ – experiment; W = 2,15 W/cm^2 (j = 0,66 mA/cm^2; U = 3300 V); P = 6 Pa, t = 300 s.

The increase of the discharge power density W to a value of 2.35 W/cm^2, respectively, a discharge current density j to values of 0.7 mA/cm^2 (Fig. 2) reduces the contact angle to 22 degrees. The subsequent increase in the values of W and j leads to growth of the contact angle. This is due to the heating surface. Therefore, visually detectable formation of the oxide film, and the growth rate of its oxidation due to the presence of oxygen in the chamber as part of the residual gas and is letting argon. For the same reason there is stabilization intensity of the treatment surface with the consequent formation of the oxide film at voltages above 2500 V. n this case, the processing efficiency deteriorates, apparently due to the fact that the rate of oxidation of the surface increases and exceeds its velocity spraying.

Since ion surface cleaning process is characterized in that the discharge current density j is determined by the pressure of residual gases in a square P^2, in the pressure range of P = 5-6 Pa also has a minimum contact angle θ = 22-24 ° (Fig. 3).

The distance between the electrodes L in the region considered the factors has a similar extreme nature. When L is less than 6 mm intensified thermal processes on the treated surface, leading to increased speed of its oxidation state (Fig. 4). The increase in the contact angle with the distances L more than 10 mm is probably due to the weakening of the frequency of ion bombardment work piece.

In these experiments discharge treatment was performed for a fixed time of 300 seconds. To determine the optimal treatment time t test was performed in the range of 30-600 s at previously determined optimal mode (W = 2,15 W/cm^2 (j = 0,66 mA/cm^2; U = 3300 V); L = 9 mm; P = 6 Pa). The results show that the minimum processing time t, ensuring optimum contact angle θ, is 180 (Fig. 5).

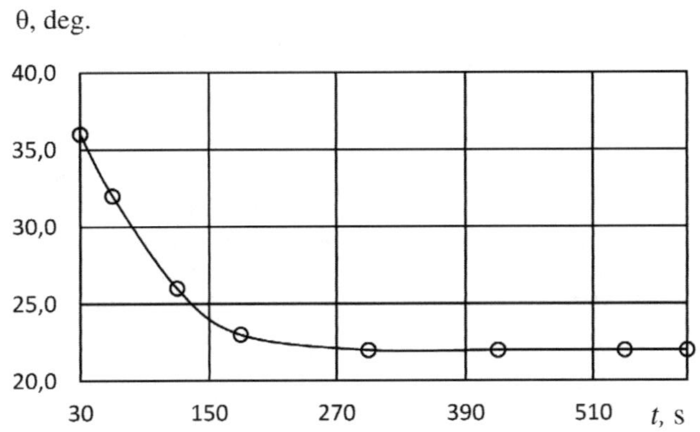

Figure 5. The dependence of the contact angle θ on the processing time t: W = 2,15 W/cm^2 (j = 0,66 mA/cm^2; U = 3300 V); L = 9 mm, P = 6 Pa.

4. Results of the study quality of cleaning the steel surface by the criterion of the bond strength of the plasma coating to the base

Effect surface treatment by a glow discharge before plasma spraying on the strength of the coating to the surface shown in Fig. 6. The results show an increase of strength between the overlay material GHG HN80SR3, PN55T45 PN85YU15 and with a core of 45 steel by 60-80% if prior to the deposition surface of the substrate treated by a glow discharge at an optimal mode. One of the classic methods of surface preparation prior to plasma spraying using an arc plasma torch is sandblasting. Therefore, the effect of treatment by a glow discharge was evaluated on the strength of the coating to the substrate after blasting. The results show for this case an additional increase in the bond strength by up to 55%.

Under real conditions, after cleaning the surface prior to the deposition of coatings in the open atmosphere, there is always a period of time during which the treated surface is contacted with atmospheric air. To assess the effect of this time on the strength of the coating to the substrate were tested, the results of which are shown in Fig. 7.

Samples were kept before spraying in air for 1, 2, 4, 10, 20, 40, 70 and 120 minutes. The nature of the changes to the bond strength of coatings on the basis of the residence time of the substrate in the air indicates the intense decreasing of chemical activity of the surface in the first minute followed by a decrease of the intensity to constant values after 40-60 minutes.

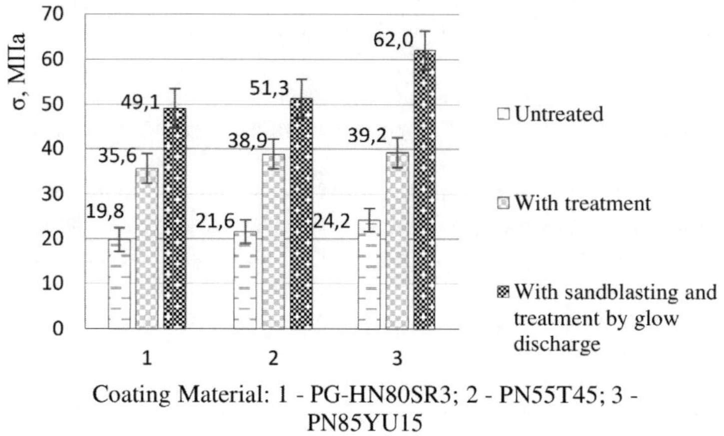

Figure 6. Effect of pre-treatment of the surface of steel 45 on the strength σ it covers

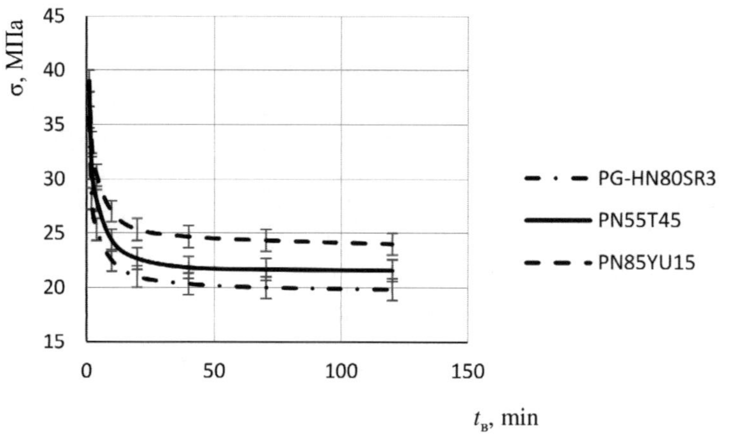

Figure 7. The dependence of the bond strength of the coating σ to the substrate of the length of the substrate in the air finding t_B between the treatment by a glow discharge and plasma spraying.

5. Conclusions

Experimental studies of the quality of cleaning the steel surface by a glow discharge have shown that it is advisable to use an abnormal discharge regime in the parameters. Specific power discharge 1,88-2,4 W/cm^2 (~ 3100-3400 V discharge voltage, the current density ~ 0,60-0,72 mA/cm^2), the residual pressure of argon in the vacuum chamber was 5,5-6,0 Pa; the distance between the electrodes was 8-9 mm; processing time – 180 s. The dependence of the criterion of quality of steel surface (contact angle) by a glow discharge from the factors of the process. This allows you to quantify the impact.

Surface treatment by a glow discharge before spraying metal and intermetallic coating enhances the strength of their connection to the base by 60% to 80% in the case of ground samples and 30-55% in the case of samples with sandblasting. Increase strength characteristics of the coatings that have passed before spraying pretreatment by a glow discharge, allows you to use the method of preparation of surfaces for the cases to provide the necessary high bond strength of coatings to the substrate. This can be actual for applications in the production of coatings, working in environments with high contact and shear loads. Obtained optimal modes treatment of steel surfaces by a glow discharge are technological recommendations of such processing.

Evaluation results of cleaning steel substrates compared with chemical etching provides the required strength quality compound at high process ecology. In comparison with the cleaning by the arc discharge – surface uniformity of treatment of simple and profiled surfaces. Further development of research to improve the efficiency of cleaning of surfaces in the direction seems dynamization mode glow discharge and the protection of the activated surfaces before applying coatings on them.

Acknowledgments

The reported study was funded by RFBR, according to the research project No. 15-38-51042 mol_nr.

References

[1] Anciferov V N, Bobrov G V, Druginin L K and etc 1987 *Powder metallurgy and sprayed coatings* (Moscow, Metallurgy) p 792

[2] Rikalin N N, Shorohov M H and Kudinov V V 1969 *On the mechanism of formation kinetics of a firm connection between the coating and the substrate during the deposition* (L, Science) P 5-28

[3] Saifutdinov A I, Saifutdinova A A, Kashapov N F and Fadeev S A 2016 *J. Phys.: Conf. Ser.* **669** 012045

[4] Saifutdinov A I, Fadeev S A, Saifutdinova A A and Kashapov N F 2015 *JETP Lett.* **102** 637–642

[5] Gabovich M D, Pleshivcev N V and Semashko N N 1986 *The beams of ions and atoms for controlled thermonuclear fusion and technological objectives* (Moscow, Energoatomizdat) p 249

[6] Bolotov G P, Bolotov M G and Rigov R N 2009 *Vestnik NTU* **(57)** 124–127

[7] Spiridonov A A 1981 *Planning experiment in technological processes* (Moscow, Mechanical engineering) p 184

[8] Tushinskii L I, Plohov A V, Tokarev A O and Sindeeb V I *Methods for Materials Research: Structure, properties and processes applying inorganic coatings* (Moscow, World) p 384

The application of additive technologies in creation a medical simulator-trainer of the human head operating field

L N Kashapov[1], N F Kashapov[1] R N Kashapov[1,2,3] and B.Y. Pashaev[3]

[1] Engineering Institute, Kazan Federal University, Kazan, 420008, Russia
[2] Laboratory of Radiation Physics, Kazan Physical-Technical Institute of the Kazan Scientific Center of the Russian Academy of Sciences, Kazan, 420029, Russia
[3] Kazan State Medical University, Kazan,420012,Russia

E-mail: kashramiln@gmail.com

Abstract. The aim of the work was to determine the possible application of additive manufacturing technology during the manufacturing process as close as possible to reality of medical simulator-trainers. In work were used some additive manufacturing technologies: selective laser sintering (SLS), fused deposition modeling (FDM), binder Jetting. As a result, a prototype of simulator-trainer of the human head operating field, which based on the CT real patient, was manufactured and conducted its tests. It was found that structure, which is obtained with the use of 3D-printers ProJet 160, most appropriate and closest to the real properties of the bone.

1. Introduction

Education and improvement of professional skill surgeons is one of the major health problems. One method of training is to practice the skills on phantoms and simulator-trainers. A simulation training improves the quality of manipulation in normal and stressful situations in anesthesiology, it allows to achieve high quality of nursing manipulation, enhances the quality of care for patient, increases the frequency of successful small surgical interventions, reduces the number of complications. Particularly important the application of simulators-trainers in the field of neurosurgery at the practice skills, as part of learning new methods of access to the brain area. Cadaver heads are the best on the realism of the existing objects at the moment. Earlier, unidentified and unclaimed bodies can be sent to medical institutions, but now prohibited their use. The body of a deceased person can be use in educational and research purposes only if his identity is established, but do not have people who are willing to bury him. To get that body, the organization shall send a corresponding request. At the moment, there is an acute demand in cadaver material. Supply of cadaver material in Russia is carried out from the United States, and it is very expensive, for example, the cadaver's head is worth $ 1,000. Also cadaver heads do not reproduce the desired clinical case for practicing quite difficult operations to remove tumors. One of the existing problems is the necessity to create a phantom-simulators for practicing methods of treatment endonasal endosurgical skull base surgery diseases, transcranial surgery of brain tumors. A special feature of generated simulators is the necessity to recreate realistic anatomy, similar mechanical and strength properties, individuality in the tumor location. The solution to this problem can be found in the application of additive technologies with the use computed tomography of real patients. The aim of the work was to determine the possible application of additive manufacturing technology in the manufacturing process as close as possible to the reality of the human head phantom to simulate the surgical field.

2. Main part

For creation of real anatomic form simulator-trainer used data from computer tomography of a young healthy man aged 21 years, obtained by X-ray tomograph SkyViev. This scanner is equipped with a conical X-ray source, wherein the slice thickness is 0.1 mm and the minimum step of longitudinal sections is 0.05 mm. The resolution of the device, a digital matrix 1000 by 1000, pixel size is 7.4 microns. The data obtained are DICOM extension. With program Slicer was designed the three-dimensional model of a human head.

After creating a virtual model of the main problem is the reconstruction of the object in a material form. It is necessary to choose the method of manufacturing and the materials that most closely coinciding with the mechanical properties of human tissues. In [1] describes a method for the manufacture of neurosurgical simulator using 3D printer Objet 500 Connex (Stratasys) that is highly effective and allows the layers to create simulators having a gradient composition and density of the material. The resulting model is very realistic, but it has a serious drawback - the high cost $ 2,600. It is very expensive and this method does not satisfy us. We have selected an approach to individual fabrication of each part and assembly.

The developed simulator-trainer should consist of several main parts: brain, tumor, dura, skull, subcutaneous tissue and skin. The skull is the main constructional support element simulator trainer, in which located brain, tumor and the dura mater covering the skull inside. And on the outside of the skull is attached to the subcutaneous tissue with the skin. Since the CT data obtained by Z-axis have a layer thickness of 100 microns, then it does not make sense to use in the manufacture of additive simulator technology to better than 100 microns. Beneath this criterion very well suited FDM-technology. For the manufacture of the skull was used 3D printer Fortus 400 mc [2,3], the construction materials produced ABS-plastic and polycarbonate. The skull was split into two parts to allow the application of the dura mater and the installation the brain with tumor. The resulting sample of skull is shown in Figure 1.

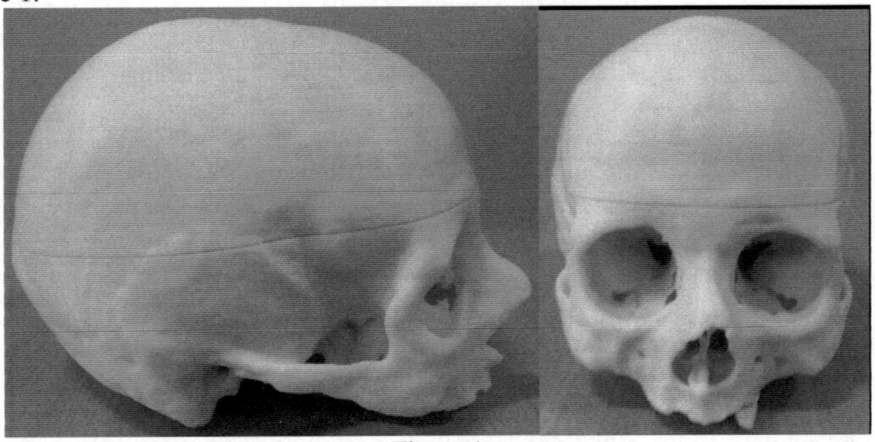

Figure 1.

The next stage is the manufacturing brain. MRI data used, which obtained by scanner General Electric Signa HDe 1.5 T. Then, using a BIOMODROID CB-MTI UM program it three-dimensional model of the brain was made in the STL format. As the mechanical properties of the brain coincides with the silicone rubber companies Bredent [3]. Shape for pouring the silicone manufactured in two parts, as shown in Figure 2.

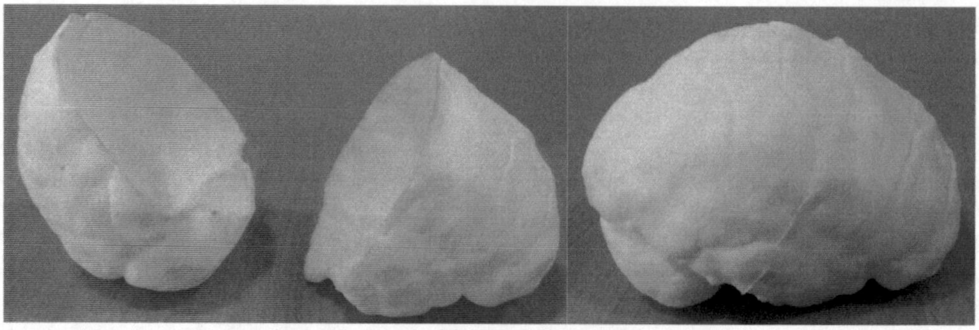

Figure 2.

The inner side of these forms is a cast of the external shape of the three-dimensional virtual model of the brain. Virtual model of forms were designed in the Geomagic software and printed on the installation of selective laser sintering SLS sPro 60 HD, from material nylon 12. This material has the following mechanical properties: tensile strength - 43 MPa; tensile modulus - 1586 MPa; relative elongation at break - 14%; strength in bending - 48 MPa; flexural modulus - 1387 MPa; Shore hardness D - 73. This material is well suited for manufacturing of forms for pouring silicone models. The tumor at the base of the skull was made similarly. The dura mater has a thickness of 0.5 to 1 mm. Liquid silicone was used for its production, which is applied to the inside of a plastic skull model, the silicone surplus flowed by gravity, and on the surface remained the silicone layer thickness of 0.8 mm, which is during 15 hours solidified.

Since in learning methods of treatment endosurgical endonasal skull base surgery diseases and transcranial surgery of brain tumors is supposed to be accessed through the nose, then the role of the subcutaneous tissue, the skin does not play a significant role and need only for fixing simulator trainer in the fixing device of the operating table. Subcutaneous tissue and skin replaced by polyurethane model shown in Figure 3.

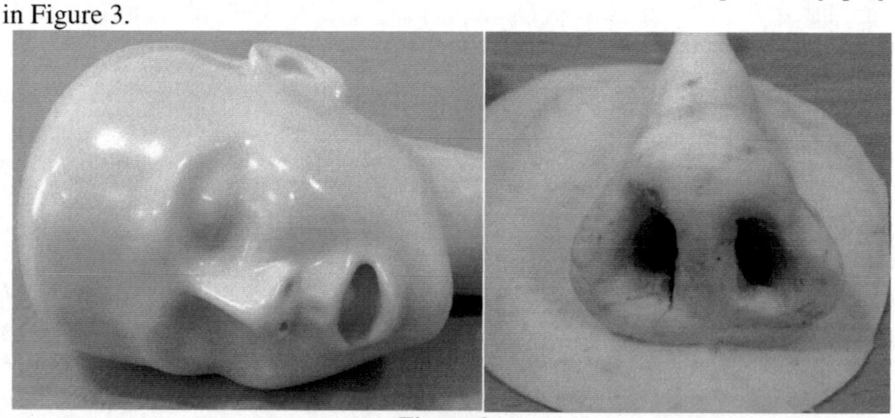

Figure 3.

While teaching on the simulator trainer surgical access should be carried out through nose, which is necessary to make a realistic, at the level of the biological prosthetics. Realized nasal region simulator trainer should have a high accuracy of fit to the skull and was made as follows. With the help of optical 3D-Scanner Range Vision, the principle of operation is based on structured illumination of the object of scanning by safe white light, and received a three-dimensional model of the nose in STL format. Further scan results of the skull and the file has been imported in the CAD-program PowerShape Pro. This program produced the combination of the nose to the skull, providing a snug fit to the skull. Virtual model of the nose has been obtained, which repeats the shape of the skull in the place of its fit. At the next stage was made master-model of the nose by using stereolithographic installation ProX 800, then master-model filled with silicone ProtoSiC RTV 245, and silicone mold manufactured for vacuum casting. Further, silicone rubber ProtoFlex 150-05 was filled in this form and obtained silicone nose for installation on the simulator-trainer (Figure 3).

Tests of a model were conducted using a high-speed motor multi-functional system UNIDRIVE SIII NEURO of STORTZ company with rotational speed 60000 rpm and DRILCUT-X II Shaver Handpice. Table 1 shows the results of the evaluation of the developed simulator-trainer by neurosurgeons of the human head operating field.

Table 1.

Area of model	The test property	evaluation	disadvantages
Skull	Anatomy Mechanical properties of ABS material Mechanical properties of polycarbonate	Good Unsatisfactorily satisfactorily	It is necessary to make the removable an insertion in the skull for multiple use
Dura mater	The possibility of cutting	Good	
The brain and tumor	formal resemblance The possibility of cutting	satisfactorily	Necessary to make vessels of the head with the ability to injection the pressurized fluid
Subcutaneous tissue and skin	The outward similarity with the head of the patient, the possibility of mounting the device of the head fixation	Good	
Nose	outward similarity	Good	It is necessary to create mucosa inside of the nasal cavity

In general, the simulator-trainer of the head of the surgical field is ideally suited for training of neurosurgeons to methods of treatment endonasal endosurgical skull base surgery transcranial diseases and surgery of brain tumors. During the tests revealed several disadvantages: necessity of creation the mucous membrane of the nasal cavity, the skull from ABS-plastic does not correspond to the real properties and melts during drilling holes, no possibility of re-manipulation of operating drilling field. These problems were solved as follows: as a material was tested polycarbonate, which has a higher melting point and strength, but this plastic wound onto surface of bur and obstructed drilling procedure while using a diamond burs. On this basis, it was decided to use thermosetting plastics instead of thermoplastics. For the possibility of reusable application the part of the skull, which is prone to destruction in the process of practice skills, was made removable, as shown in Figure 4. Because of the need to use thermosetting plastics this replacement part manufactured on ProJet printer. The carcass of insertion impregnated by special binder based on polyester resins after receiving.

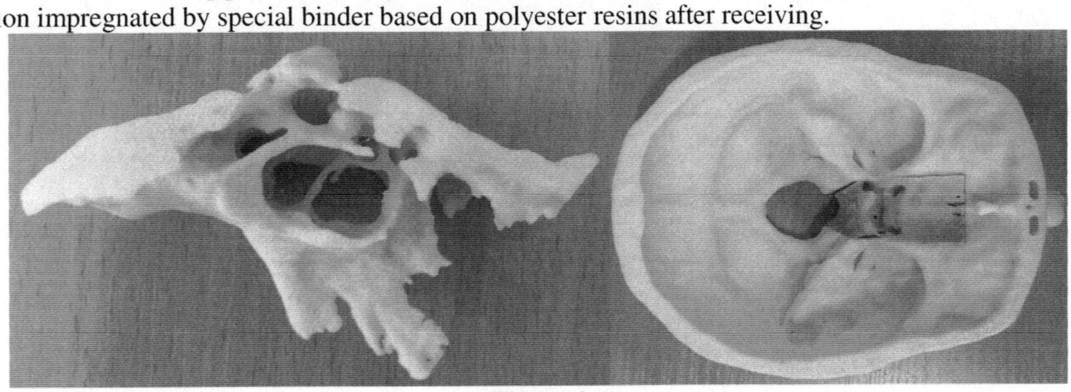

Figure 4.

3. Conclusions

The developed method of manufacturing simulator-trainers of the head surgical field with the individual characteristics (location of the tumor, the individual size, etc.) is very convenient and economically feasible in the manufacture of small quantities up to 20 pieces. This is due with the fact that on average, a trained team to practice manual skills consists of no more than 20 people. Interchangeable insertion of allow on the same simulation increasing the number of cycles of repetition thereby achieve the transition effect of "quantity" to "quality".

Existing at the moment 3D-printed simulators worth $ 2600 [1], produced by our method was $ 200. In the future we plan to improve simulator-trainers and to develop a method of manufacturing the circulatory system to filing fluid under pressure.

This work was funded by the subsidy allocated to Kazan Federal University for the state assignment in the sphere of scientific activities.

References

1) V Waran, V Narayanan, R Karuppiah, S Owen, T Aziz, J Neurosurg 2014 120 pp 489– 492
2) L Kashapov, A Rudyk, R Kashapov 2014 *IOP Conference Series: Materials Science and Engineering* **69** 012016
3) R Kashapov, A Korobkina, E Platonov, G Saleeva 2014 *IOP Conf. Series: Materials Science and* Engineering **69** 012026

Innovative production technology ethanol from sweet sorghum

N F Kashapov[1] M M Nafikov[1] M X Gazetdinov[2] M M Nafikova[2] and A R Nigmatzyanov[3]
[1]Kazan (Volga region) federal university, Kazan, Russia
[2]Kazan state agrarian university, Kazan, Russia
[3]Tatar institute of retraining specialists agribusiness, Kazan, Russia

Arnig76@yandex.ru

Abstract. The paper considers the technological aspects of production of ethanol from non-traditional for Russian Federation crops – sweet sorghum. Presents the technological scheme of alcohol production and fuel pellets from sweet sorghum. Special attention is paid to assessing the efficiency of alcohol production from sweet sorghum. The described advantage of sugar content in stem juice of sweet sorghum compared with other raw materials. Allegedly, the use of the technology for producing alcohol from sweet sorghum allows to save resources.

The main criterion in the production of alcohol is its cost: the lower cost of production, so it is more competitive. The cost of alcohol is greatly influenced by raw materials and production technology. In our country food alcohol produced from grain and potatoes, and technical - from sawdust [1, 2, 3, 4, 5].

The cheapest alcohol currently produced in Brazil from sugar cane. This is the most simple and cheap technology of production of both sugar and alcohol. It consists of the following cycles: clean the stems from the leaves, pressing the stems with the release of the juice, adding the juice of yeast and the fermentation of sugars to alcohol, separating the alcohol by distillation.

For alcohol, you can use any materials containing significant amounts of sugar or materials that can be converted into sugar, for example, starch or cellulose.

In Russia the climate in most part of the territory dictates the cultivation of a limited number of plants, which may receive sugar. In addition to traditional crops, there is a possibility of cultivation of sweet sorghum to obtain ethyl alcohol (table. 1).

Table 1 - Efficiency of alcohol production from different crops

Raw	Production from 1 ha, m3	Cost, USD / m3
Sugar beet	2,5-3,0	300-400
Sugar cane	3,5-5	160
Corn	2,5	250-400
Wheat	0,5-2	380-400
Potatoes	1,2-2,7	800-900
Sugar sorghum	3,0-5,0	200-300

Calculations show that the lowest cost is alcohol from sugar - and starch-containing products such as sugar cane, sweet sorghum and corn.

With regard to involvement in agricultural turnover of raw land in Russia may be expanded in crops of sorghum and maize to 18-20 million hectares, which will produce by 2020 about 1.5 million tons of alcohol without breaking the rotation (Fig. 1).

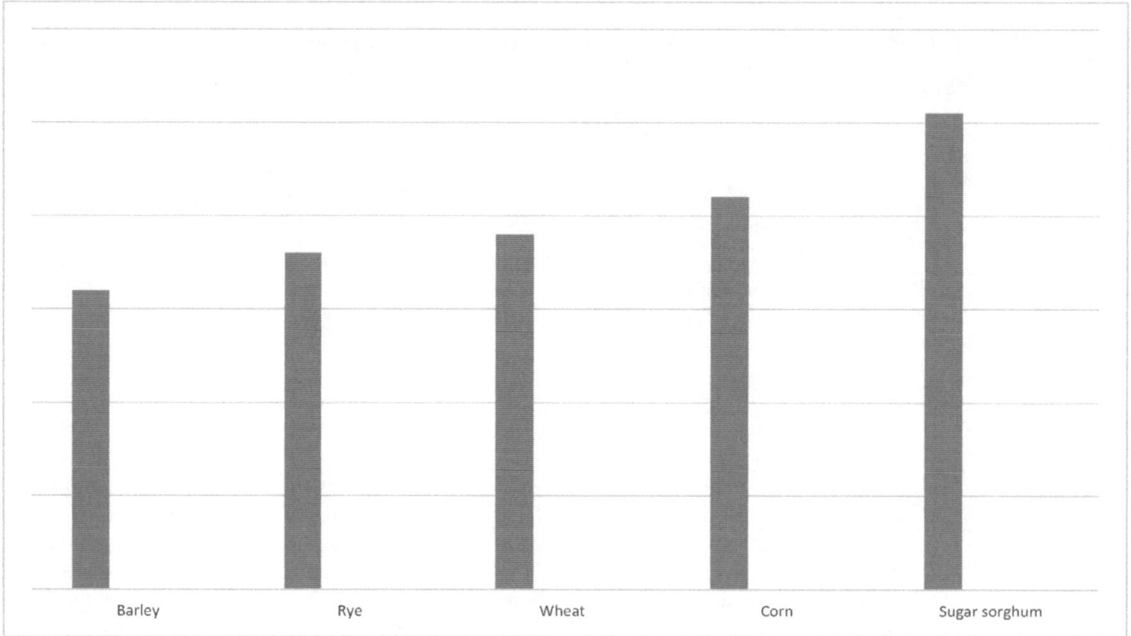

Figure 1 - The yield of alcohol from various raw materials (at 90% conversion of starch into alcohol)

Sweet sorghum in the world agriculture occupies more than 30 million hectares In Russia, sorghum is cultivated on an area of over 700 hectares. Culture is common in the southern regions of the European part (Rostov region, Stavropol and Krasnodar Krai), Lower and Middle Volga region, the Urals and the far East. Exceptional drought tolerance and high yields make this crop very promising for cultivation in the arid regions of the Russian Federation, including the Republic of Tatarstan [6, 7, 8, 9,10,11].

The content of sugars in the stem juice of sweet sorghum is not inferior to the sugar cane, but the juice differs in composition. If sugar cane in its juice contains only sucrose (crystallizing sugar), the juice of sweet sorghum contains sucrose in addition to the largely glucose and soluble starch, which interferes with crystallization. In addition, an important characteristic is the fact that sweet sorghum belongs to the plants with C4 photosynthesis and therefore photosynthetic potential is 2-3 times higher than that of spring wheat, soybeans and sugar beets. The maximum number of accumulation of sugars coincides with the phase of wax and full ripeness of grain. From one hectare of sweet sorghum sowing you can get 5-7 tonnes of sugar and 15 tons of pulp (dry bagasse), and 3-5 m3 of alcohol. We have proposed the technology of cultivation of sugar sorghum and the scheme of obtaining from plant raw material alcohol and fuel pellets.

Process flow diagram of alcohol production and fuel pellets are presented in figure 2.

Received from the field, the crop material is crushed to facilitate the further deflection of his juice. Then it enters the mill. The mixture should be pressed twice, moistening between the first and second time with water to dilute the pulp contained in sweet liquid (maceration). The maximum possible squeeze 90-93% of the total contained in the stems of sugar with a small diluted juice. Extracted juice

IOP Conf. Series: Materials Science and Engineering **134** (2016) 012012 doi:10.1088/1757-899X/134/1/012012

passes through the filter, where impurities are separated, small pieces of stems, etc. pressed Solid fraction is used for production of solid biofuels (pellets).Part of the pressed juice is taken and fed into projektering for the cultivation of yeast, the remainder is mixed with yeast in a fermentation apparatus.

For fermentation in a vegetable juice add to the culture of Saccharomyces cerevisiae with enzymes (invertase and simaz) process the sucrose into ethanol.

$C_{12}H_{22}O_{11} + H_2O = C_6H_{12}O_6 + C_6H_{12}O_6$

sucrose glucose fructose

zimasa

$C_6H_{12}O_6 = 2C_2H_5OH + 2CO_2$

Glucose (fructose)/ ethanol

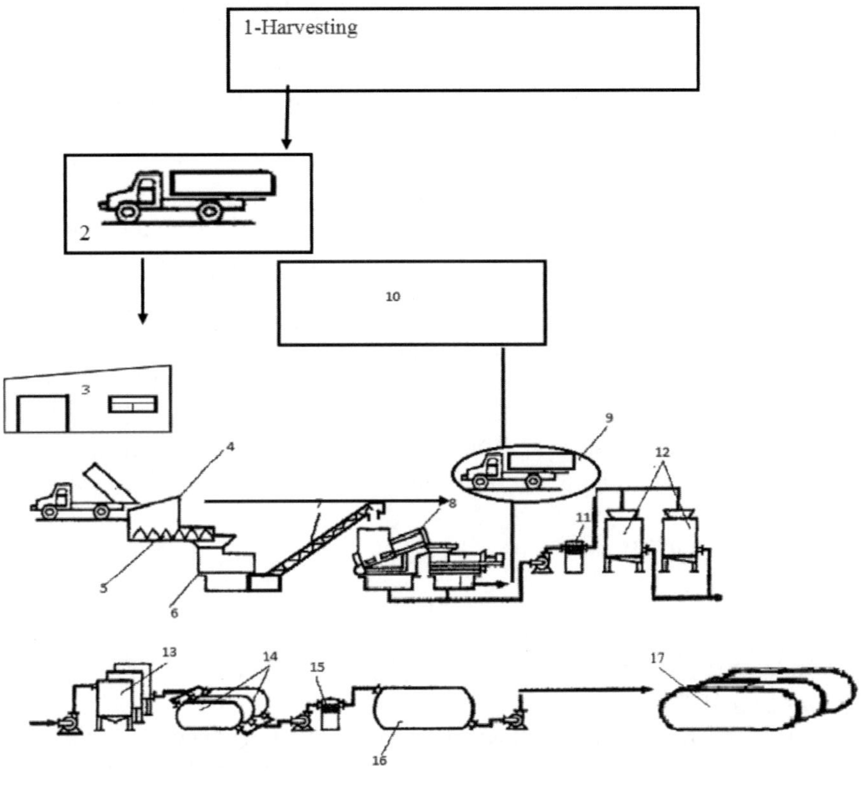

1. Harvesting 2. Transportation truck 3. The weighing machines, laboratory 4. Receiving hopper	5. Belt conveyor 6. Shredder 7. Screw conveyor 8. Press 9. Transportation truck	10. The trench hay(pellets) 11. The coarse filter 12. The capacity for digestion	13. The capacity for maturation of the finished product 14. The evaporator 15. The filter of thin clearing 16. Distribuye 17. Storage

Figure 2 - Scheme of the technological process of production of alcohol and fuel pellets from sweet sorghum

Well rabble the product the ethanol content shall not be less than 10%, but not more than 13%. The concentration of residual sugars - not more than 0.45%, the acidity does not exceed 0.2%. On the ethanol yield, the rate of its formation and the physiological state of the culture Saccharomyces cerevisiae is greatly influenced by their concentration in the fermentation product. The optimum concentration is 15-20 g per 1 l of the fermentation product. This amount of yeast spaevaet sugar quickly, usually within 6-7 hours, and the ethanol yield reaches 57-59 l of 100 kg fermentable sugars. Next, the rabble, the product is distilled, the method is based on the fact that ethanol and water have different boiling points. If the water under normal conditions boils at a temperature of 100°C, then ethyl ethanol need to 78°C. in Addition to water and ethanol rabble product contains a lot of impurities. The lungs begin to boil at a temperature of 68°C, and heavy – 85°C. therefore, the distillation of the fermentation product, it is important to achieve such a temperature at which the ethanol is boiling, fusel oils yet. At the outlet of a distillation column ethanol is obtained by a fortress no more than 96,2.

Work on the production of alcohol is carried out in 3 shift of 8 hours. To work on presents a technological line of manufacture of alcohol requires 3 main employee - operator, to operate equipment required 2 auxiliary worker (mechanic, electrician).

Conclusion

1. World and domestic experience show the great economic value of sweet sorghum as culture comprehensive use of. It is a highly valuable feed and industrial crop. In the vegetable juice of sweet sorghum contains sucrose in addition to the largely glucose and soluble starch, which interferes with crystallization.

2. In the Republic of Tatarstan the possibility of cultivation and use of crop sweet sorghum for sugar to 3 to 5 tons and eco-friendly food alcohol to 3-5 m3 per 1 ha of sowing.

References

[1] B A Alabushev, L N Anipenko, N G Gursky, N I Kolomiets, P I Kostylev, P A Mangush, O I Alabushevo 2003 *Sorghum (breeding, seed production, technology, economics)* – Rostov-on-Don: ZAO Book 368 p

[2] V A Alabushev, S I Gorpinchenko 2008 Use of sorghum for bioethanol production *Corn and sorghum* 5 pp 7-12

[3] T M Atamanenko, V A Maltsev, A I Rzhevsky 1993 Sweet juice for cows *Corn and sorghum* 1 pp 10-11

[4] M P Zhukov, A N Abaldo 2004 The Breeding of sorghum crops in the Stavropol select entry *Corn and sorghum* 5 pp 14-17

[5] Nigmatzyanov A R 2008 The Creation of enterprises for production of biofuel made from potatoes *Issues of improving the mechanism of management of economic systems on the food market:a Collection of articles all-Russian scientific Internet conference with international participation*, Kazan: Kazan state University 653 p

[6] V Gusev, VV Larina, K A Kulikova, T Y Nikitin 2004 Breeding of sorghum in niiskh of the South-East *Corn and sorghum* 3 pp 22-24

[7] Y I Isakov, S I Gorpinchenko 2003 Selection of sweet sorghum *Corn and sorghum* 1 pp 9-12

[8] Nafikov M M 2012 The dependence of the yield of sugar sorghum from the methods of seedbed preparation *Corn and sorghum* 3 pp 21-24

[9] Fomin V N, Nafikov M M, Valiev I Z 2012 The yield of sorghum depending on the ways of main soil tillage and fertilizers in conditions of forest-steppe of the Volga region *Advances in science and technology AIC* 2 pp 25-27

[10] Kashapov L, Kashapov N and Kashapov R 2013 *J. Phys.: Conf. Ser.* **479** 012005

[11] Timofeeva O A Shaymullina G H and Nevmerzhitskaya J Y 2015 Dynamics and Polypeptide Composition of Lectins in Wheat Germs Infected With Plant Pathogenic Fungi *Research Journal of Pharmaceutical, Biological and Chemical Sciences* 6(4) pp 2190-96

Justification of the choice of units for mains-noah soil cultivation of sweet sorghum and their effectiveness

N F Kashapov[1] M M Nafikov[1] M X Gazetdinov[2] M M Nafikova[2] and A R Nigmatzyanov[3]
[1]Kazan (Volga region) federal university, Kazan, Russia
[2]Kazan state agrarian university, Kazan, Russia
[3]Tatar institute of retraining specialists agribusiness, Kazan, Russia

e-mail: Arnig76@yandex.ru

Abstract. The article is devoted to problems of improving the efficiency of tillage crops. Presents an approach that focuses on the application of resource-saving technologies. To investigate the relationship between the financial welfare of the management and selection of units for primary processing of the soil. Conducted economic evaluation and identified the energy efficiency of main processing of the soil under sweet sorghum.

A set of methods of mechanical impact on the soil before planting and during plant growth is called a system of soil treatment. The basic processing of soil are of great importance since it largely affects the water-physical, biological-physical and agrochemical properties of the soil, which in combination with other techniques ultimately determines the magnitude of crop yields. When processing of the soil improves the structure of the topsoil and as a result of tillage increases the total porosity and the reduction is the amount of capillary pores. Capillary decreases and increases convectiondiffusion evaporation of moisture, and also improves the heating of the soil. Conversely, compaction total soil porosity decreases, capillarity increases. In other words, the treatment of the soil creates favorable conditions for the inflow of moisture from the lower layers to the upper and reduction of aeration, which affects the intensity of aerobic processes.

Therefore, regulating the degree of compaction of the soil by means of processing, can in considerable extent, influence the preservation and accumulation of moisture, and the conditions of life-activity of the soil microflora. In very loose soil (usually this happens immediately after processing) mineralization of humus increases and decreases its content. When more raft by the addition of mineralization is reduced and created the best conditions for humification of organizational-technical substances. Loose topsoil that is created after processing, under the influence of gravity, precipitation and other factors samoletnaya to so-called equal-spring density.

Direct assignment technology in agricultural production is the execution of technological operations in the cultivation of crops that provide their biological-ecological requirements. As a result, should get cost-effective and Vysokyi prominent, environmentally friendly products [1, 2, 3].

At the present stage in the agricultural sector use a wide range of farm machinery Noah. In this regard, we have conducted field experiments on the study of various tools for primary soil cultivation of sweet sorghum and their effectiveness in the conditions of the Republic of Tatarstan.

To soil could contribute to the generation of possibly large harvests necessary to possess a certain structure. Bulk density of the soil in the SIG-nificant extent, shape water, air and nutrient regimes, as well as the activity of soil biota. High bulk density of the soil increases the development denitrificatori and butyric acid bacteria prevents the accumulation in soil of mobile forms of nitrogen, phosphorus and potassium.

In our studies, in addition the density of the soil before sowing sorghum in the variants of tillage significant differences were not, however, before harvesting in the 10-30 cm layer of the soil was denser on the variants in the handling of APC-6, KPI-3.8 and CTS-10. (tab. 1).

Table 1 – Bulk density of the soil depending on processing systems, g/cm3

Basic soil	Before sowing				Before harvesting			
	0-10 cm	10-20 cm	20-30 cm	0-30 cm	0-10 cm	10-20 cm	20-30 cm	0-30 cm
Moldboard plowing PN-5-35	1,06	1,12	1,14	1,11	1,08	1,15	1,17	1,13
Plowing plows Maltseva	1,07	1,13	1,16	1,12	1,10	1,17	1.20	1,16
Treatment of APC-6	1,09	1,16	1.20	1,15	1,14	1,22	1,25	1,20
Processing KPI-3.8	1,08	1,14	1,18	1,13	1,12	1,20	1,23	1,18
Processing CTS-10	1,08	1,15	1,19	1,14	1,13	1,21	1,24	1,19

The tests showed that the hardness of the soil layers depended on the tillage systems and the degree of hydration (tab. 2). Table 2 shows that in the layer of 0-15 cm between the systems processing this indicator is not significantly different, and at a depth of 20 cm and lower it was higher on plots loosening by a subsurface cultivator.

Table 2 – Hardness of the soil depending on tillage systems g/cm3

Basic soil	At the beginning of the growing season					Before harvesting				
	5 cm	10 cm	15 cm	20 cm	25 cm	5 cm	10 cm	15 cm	20 cm	25 cm
Moldboard plowing PN-5-35	2,9	5,2	14,1	17,1	30,2	12,1	18,5	27,3	32,1	33,8
Plowing plows Maltseva	3,0	5,3	14,4	18,0	30,3	12,6	19,3	28,2	33,5	35,1
Treatment of APC-6	3,2	5,7	16,5	22,1	31,6	13,4	22,3	34,1	37,6	41,4
Processing KPI-3.8	3,0	5,4	15,6	20,1	30,5	12,9	21,2	31,8	36,3	38,2
Processing CTS-10	3,1	5,5	15,8	20,3	30,7	13,0	21,4	32,9	36,7	40,1

In experiments conducted systematic observation of the status of the nutrient regime of the soil. Three times during the growing season for three years on two layers of the arable layer in the two-times repeat activity conducted analyses to determine the content of each of the main elements of Pieting – N, P2O5 and K20 in the soil, we have derived (15 definitions) average for 3 years by layers of 0-15, 15-30 and 0-30 cm In the result of the APC-6, KPI-3,8, CTS-10 under crops is clearly seen the differentiation of the parts of the arable layer on fertility. In the layer 0-15 cm, the content of P2O5 on the moldboard treatment was 164 mg/kg, with processing APC-6 – 176 and CTS-10 – 173 mg/kg, K2O respectively 179,186 and 184 mg/kg. Excess options APC-6 P2O5 – 12, K20 – 15 mg. of Alkali-hydrolyzable nitrogen by plowing contained 78, and when the soil treatment unit LDC-10 – 85 mg/kg of soil. In the underlying layer (15-30 cm) Pro-a pattern opposite signs. Plowing found 71 mg, and tillage CTS-10 – 55, P2O5 150 and 133, respectively, K2O – 161 and 150 mg/kg.

Therefore, the fertilizer applied, particularly phosphate and potash, when the flat by loosening trapped mainly in the upper (0-15 cm) layer and in lower layer (15-30 cm) penetrate in much smaller quantities than is the case in plow treatment.

The results of the research identified economic efficiency, which are presented in table 3.

Table 3 - Economic evaluation of the use of different tools for primary soil cultivation of sweet sorghum

Option experience	The yield of green mass, t/ha	The cost of the gross commodity, rub.	Tomorrow-are you on 1 ha, rub.	Net income from 1 ha, rub.	Profitability level, %	The cost of 1 ton of green mass, rub.
Moldboard plowing PN-5-35	41,1	5014	3050	1964	64,3	74,2
Plowing plows Maltseva	35,9	4380	2954	1426	48,3	82,2
Treatment of APC-6	26,5	3233	2850	383	13,4	107,5
Processing KPI-3.8	31,4	3831	2872	959	33,4	91,6
Processing CTS-10	30,6	3733	2865	868	30,3	93,6

The greatest economic efficiency revealed in the cultivation of sweet sorghum in the application of moldboard plowing [4,5]. Net income in this embodiment, 1964 amounted to rub/ha, profitability level 64.3 per cent, the coefficient of energy efficiency of 6.7 vs. 383 - 1426 RUB./ha , the 13.4 - 48.3 percent and 4.4 to 5.7 obtained in other ways. Bioenergy until the indicator of agro-ecosystem in these variants was respectively 4.5 and 1.1.

However, in recent years more and more importance in international practice acquires bioener-energy evaluation[6, 7]. It allows to compare different technologies of production of agricultural products (tab. 4).

Table 4 – Energy efficiency primary tillage in the cultivation of sorghum with various tools

Option experience	The yield of green mass, t/ha	Accumulated energy, GJ/ha	Energy, GJ/ha	Energy efficiency ratio	The PSU agroecosystem
Moldboard plowing PN-5-35	41,1	189,88	30,23	6,3	4,5
Plowing plows Maltseva	35,9	165,86	29,11	5,7	2,6
Treatment of APC-6	26,5	122,43	28,01	4,4	1,1
Processing KPI-3.8	31,4	145,07	27,91	5,2	1,7
Processing CTS-10	30,6	141,37	28,03	5,0	1,6

The highest (6,3) in the experiment the coefficient of energy efficiency obtained about the development of a soil as well plough PN-5-35, the lowest (4,4) – when tillage Agra-gat APC-6. Bioenergetic indicator of agroecosystem in these embodiments, amounted respectively to 4.5 and 1.1.

Conclusion

1. The lowest (16-24 PCs/m2) weediness was in the variant with conventional tillage and highest when conducting primary tillage aggregates KPI-3.8 (49-55 PCs./m2) and CTS-10 (49 to 56 pieces/m2).

2. When tillage plows Maltseva and KPI-3.8 before planting sweet sorghum accumulates more moisture than when plowing. The difference in favor of KPI-3.8, compared with tillage by year 4-5 mm. To harvest the moisture content is aligned from the greater clogged-ness of crops and the deterioration of the physical properties of the arable layer in other embodiments.

3. At the option of moldboard plowing in three years on the average of the maximum (of 41.1 t/ha) yield of sweet sorghum. Below are some (of 35.9 t/ha) yield of sorghum was under obra-design soil plows Maltsev. The lowest (of 26.5 t/ha) productivity of green mass of sorghum-La when the tillage unit of APC-6. When tillage KPI-3.8 and the CTS-10, it amounted to Vila respectively 31.4 and 30.6 t/ha.

4. Taking into account the multifactorial formation and to improve the financial blah-popoluca agricultural enterprises should implement measures to validate the aggregates of primary tillage for each crop taking into account biological, economic, and technological factors.

5. Under sweet sorghum cost-effective and energy efficient were ploughing on-lesnym the plow PN-5-35.

References

[1] G I Kazakov and V A Korchagin 2009 Conservation tillage are, On Average, Volge *Crop* 1 pp 26-28

[2]. Kalichkin V K 2008 Minimum tillage in Siberia: problems and per-prospects *Agriculture* 5 pp 24-26

[3]. Mazitov N T, Garipov N E, Sakhapov R A 2007. Domestic competitive resource-saving technologies of tillage, seeding and harvesting promising units *Niva of Tatarstan* 1 pp 36-37

[4]. Nafikov M M 2012 The effect of methods of main soil tillage on the productivity of sorghum in the conditions of forest-steppe of the Volga region *Corn and sorghum* 4 pp 8-11

[5]. M M Nafikov, I Z Valiev, V N Fomin 2013 Assessment of some elements of agro-techniques of sorghum in the conditions of the Kama river region of Tatarstan forage production 2 pp 28-29

[6]. Bazarov E I, Glinka M I, Kayumov M K 1983 Energy assessment efficiency of agricultural production *M.: agricultural Sciences* 56 p

[7]. Balaur N and Aunt B 1983 Application of energy analysis to assess the efficiency of technologies of cultivation of field crops *Chisinau: mold. NII STI* 23 p

[8]. Kashapov L N, Kashapov N F and Kashapov R N 2013 *J. Phys.: Conf. Ser.* **479** 012003

[9]. Timofeeva O A, Khusnetdinova L Z, Shaymullina G H, Nevmerzhitskaya J Y 2015 *Research Journal of Pharmaceutical, Biological and Chemical Sciences* 6(4) pp 128-133

IOP Conf. Series: Materials Science and Engineering **134** (2016) 012014 doi:10.1088/1757-899X/134/1/012014

Economic efficiency and effectiveness of ways of separating materials electro diamond processing

I I Khafizov
[1]Kazan (Volga region) federal university, Kazan, Russia

Abstract.
Purveying operations on the division of all types of materials include a hand and machine scission on the equipment of the different setting. In an engineer there is reliable information about the mastered methods, their maximum possibilities and defects. With the increase of stake of expenses there was a problem of research of new types of division of materials on materials, especially it touched scarce and expensive alloys.

Introduction

Harvesting and separation operation of all types of materials include hand and machine cutting equipment for various purposes. For this purpose, traditional methods (press, metal-cutting equipment with metal and abrasive tools, etc.) and new types of processing (laser cutting, EDM division, ultrasonic processes). In engineering there is reliable information about mastered the methods, their limits and shortcomings. With the increase in the share of material costs was an issue of finding new types of separation of materials, especially of scarce and expensive alloys of the precious metals, tungsten, magnetic alloys, fragile semiconductors, where the yield details after processing was less than half of the initial mass, and the defects introduced into the surface layer when cutting, is preserved in the product and reduced its performance. In accordance with the definitions of the national standard GOST R ISO 9001-2015 performance refers to the degree of implementation of planned activities and achieving planned results and effectiveness - the link between the reached result and used resources. As the object of study selected two groups of dielectric materials: ductile and brittle, have high cost and very scarce. In world practice, the separation of such materials are reinforced disks, EDM, ultrasonic method, a laser. However, such methods are not capable of providing the required high requirements on precision, surface layer quality, performance, lower wastage of material. In addition, most methods causing environmental pollution (dust when using abrasive, etc.)

The combination of different effects on object processing allows to design a combined methods, in particular electroablation (electrolysis). These methods are applied for cutting in the production of billets with subsequent treatment, which in some cases (the manufacture of parts of instruments, radio, controls) undesirable, because it leads to undue loss of material, secondary errors and defects, and dramatically increases the cost of products. The establishment of unequivocal relationships between the properties of the processed materials, the combination of the impacts of the combined process allows you to create modern automated equipment with the office of mechanical, chemical, erosive component in a single process, providing after separation of materials finished parts with an accuracy of above 30 μm and a roughness not higher than 0.32 μm. This eliminates the negative impact on the environment and is accelerated to 2 times the cycle of manufacturing parts.

The use of these processes accelerates the creation of new competitive products, expanding the technological capabilities of production, helping to reduce the deficit and the cost of materials. It is important for modern engineering and to meet global requirements for new products.

1. The theoretical part

There are some circuits of low-waste separation of materials with the imposition of an electric field. 1.Electrochemical machining with stationary electrode. Under this scheme share of thin sheet material, applies information (serial numbers, product codes, etc.), remove burrs, sharp edges scroogled. Electrode-the tool does not move to the treated surface — electrode gap as the removal of metal from the workpiece increases, and the rate of flow of the electrolyte decreases. The process will be transient with non-stationary on the time-Meaney mode of processing. This greatly complicates the calculations of technological parameters regulation and control of the process.

In addition, the separation process does not allow to cut parts of a thickness exceeding 0.5 mm (single cutting). While processing errors increase with increasing thickness of the workpiece and make the cut with this method of round billets not promising.

2. Known method of separation of materials by ink jet method. Electrode-the tool consists of a busbar, washed by the flow of electrolyte.

Current lead is located inside the housing of insulating material. The electrolyte creates a conductive channel between the current supply and the workpiece. At the site of contact of the liquid with the processing surface of the workpiece material is dissolved and forms a recess. As it took inclusion the hole depth casing together with the workpiece. The process is fast enough only at high voltages (several hundred volts). So cut out the contours of complex-shaped parts. This method is not suitable for the separation of materials with thickness more than 0.3 mm.

1. The cutting profile or tool apropriirovannye involves separating the billet into pieces - cutting and getting the non-rectilinear contour - cut, which runs only apropriirovannye electrode-tool. Profile of the electrode-tool when cutting of parts may be made in the form of a disc or plate. His move to the workpiece at a speed and in the plane of rotation along the part. Processing is performed in a bath of dielectric fluid. If the cutting plate is performed with a single translational movement to the workpiece, it will be flashing.

In the case of apropriirovannye electrode tool perform in the form of a round wire with a diameter of 0.02 to 0.3 mm or a rod that can move in different directions at a speed and in any part of the workpiece. To eliminate the effect of wear of the electrode tool on the accuracy of the slots cut by a wire or rod is moved (usually by peremanivanie) along the axis with speed. Perform cutting in a bath of dielectric fluid[1]. Combined methods of treatment are aimed at intensification of the process of anodic dissolution. The rate of metal removal and accuracy of shaping in electrochemical processing depends on how quickly the reaction of the transition of the workpiece material in the sludge. The rate of anodic dissolution is limited by the presence of the film, passivating the surface, and the thickness of the diffusion layer, which overcome the removed treatment products.

When electroerosive grinding of solid particles (abrasive grain or filler) remove the tape, activating the process of electrochemical machining. The size of abrasive grains determining the spacing between the electrodes, as a rule, does not exceed tenths of a millimeter. At such small gaps, the current density will be substantially more than in the case of dimensional electrochemical machining. Dramatically increases the speed of metal removal in the area of the abrasive grains of the tool. In addition, some allowance of prowess is mechanical grinding. Unlike normal grinding, when the anode-abrasive processing on the surface of the workpiece is not formed stronger work-hardened layer, and manufacturer of grinding increases. Consequently, the removal rate of the metal during anodic dissolution increases due to the mechanical removal of passivating film and accelerate the process of removal of products of processing from the gap, and electrochemical dissolution of metal parts, in turn, can increase the speed of the mechanical grinding. In addition to these components of a pickup at small gaps can take place EDM process. At small sizes of the gap portion of the metal billet is removed by electrical erosion[2]. When electro-abrasive polishing allowance is removed by anodic dissolution of the metal and removing the abrasive grain or dissolution. In the first case, the tool contains a linked or free abrasive powder in the second — as a tool use a wooden or plastic bars that are located between the metal electrodes-tools. Comparing the performance of different methods, you can determine the most effective use in engineering.

IOP Conf. Series: Materials Science and Engineering **134** (2016) 012014 doi:10.1088/1757-899X/134/1/012014

Electrical discharge machining in electric mode, occurs at relatively low energy pulses. The volume of metal remote for each pulse is small, and the depth of the hole is negligible. This mode allows to obtain surfaces with high precision and small roughness at low productivity. In addition, the process is very energy intensive. Estimate the intensity ratio of electric power consumption to the mass removed from the workpiece metal. The energy intensity in the processing in spark mode much higher compared to machining PA similar operations. Great wear profile of the tool. With that said processing in spark mode effective for the manufacture of precision parts with small dimensions. Efficiency is further enhanced if the material of the part is difficult the traditional governmental methods of mechanical treatment or if the treated surface has a complex shape. Such details are characteristic for instrument making, exact mechanical engineering, tool Pro-duction of.

Processing in electric-pulse mode is characterized by greater energy level - the height of the irregularities. But due to the increase of the discharge energy, a high productivity process, which in 15 ... 20 times higher than in spark mode and is for steel of up to 250 mm3/s, for hard alloys is 3 ... 5 mm3/s. Given the small wear of the electrode tool and satisfactory energy not exceeding the same indicator for milling, treatment on electric-pulse mode can be recommended for replacing the milling of large cavities of complex shapes, recesses, channels where machining is unable to achieve high performance or where access to the tool in the cutting zone. These products are used in many industries, particularly in energy and transport machine-building, engine building, electronic industry.

Electrocontact cutting in the liquid allows to obtain process performance up to 400 ... 450 mm3/s, which is considerably higher than during the mechanical cutting of preforms. However, surface finish and machining accuracy are low. The method is economical - the consumption of electricity in 6 ... 10 times lower than in the processing in spark mode. Significant wear of the electrode tool and inconvenient in operation, the working fluid that is sprayed. This causes pollution of the machines, parts, clothes working, and requires special structures miscellaneous baths.

Electrocontact cutting in the liquid is used as a blanking operation upon receipt of the blanks from hard conductive materials.

The cutting profile of the tool electrode in spark mode allows you to perform:

• the grooves and cracks of small width, for example in the collets;
• profile electrodes-tools for flashing;
• cutting workpieces made of tungsten and other materials.

Apropriirovannye cutting electrode-tool get:

• narrow through or blind slit;
• punches with small dimensions made of hard alloys and hardened steels;
• the working part of cutters and other tools;
• slits in the collets for work pieces of small diameter (less than 2 ... 3 mm);
• tablets of the magnetic tungsten alloys, the processing of which is required to achieve a minimum consumption of material;
• labels, signs, prints.

Electrocontact cutting of the disk or tape used in the liquid:

• to receive workpieces of hard materials (stainless, HRSA, titanium, magnetic, and other alloy-WWII);
• for cutting pipes made of construction materials;
• for cutting narrow slots and holes in parts.

Cutting in air is widely used in metallurgical industry for:

• separation of part of the rental;
• cutting the sprues with large castings.

2. Systematization of modes separation of materials with electrical fields
For more complete advantage of electrochemical machining is necessary to design parts with consideration of the peculiarities of the process of anodic dissolution of alloys. It should be considered that when electrochemical machining there is no division into roughing and finishing operations - any mode electrochemical machining height of the irregularities corresponds to finish machining operations, and with increasing the metal removal rate the surface roughness decreases. Unlike machining technological parameters of electrochemical processing even increased with increasing hardness of the workpiece material. In addition, when the electrochemical processing tool or do not wear, or wear not-significantly (for the combined method of treatment).
Dimensional electrochemical machining significantly expands the technological capabilities of manufacturing parts. It is possible to obtain the surface shape that other methods or impossible, or unprofitable.

According to the scheme of cutting can be obtained with high precision delicate parts without deformation and Burr. The electrode-tool made in the form of a thin disk. The anodic dissolution takes place without significant efforts on the workpiece so it does not deform the workpiece can have a rotary motion that accelerates the process of cutting. The rate of deepening of the tool can reach 0.2..0.3 mm/s, the error processing is within 0.1 mm. When it reached the edge fillet with a radius of 0.5 mm. When using apropriirovannye electrode wire is possible to obtain grooves of complex shape; the straight sections are paired as the radii R1, R2, R3, R4, and practically radius (R=O). The cutting speed can reach 0.2 mm/s, the width of the groove is 0.5 .. 5 mm, depth of 0.1 to 20 mm, the error of width up to 0.1 mm.
3. Analysis of the methods of separation of materials
Apply methods of separation of metals let, mainly to perform blanking operations, which do not require high accuracy and quality surface layer, which are provided on subsequent stages of processing that require significant allowances a process with high complexity and energy consumption. Electrical discharge machining apropriirovannye electrode provides the accuracy of the process, but requires subsequent operations to achieve the required purity. In addition it is much more time-consuming compared with the reinforced cutting disc, which makes it not efficient for mass production. The division reinforced disc increases on the order of and higher loss of material and ensures stable performance on the accuracy of the cut, which causes the need for finishing operations. The known design of disk tool to give a greater separation width of a groove or not provide the required accuracy of the process, the quality of the surface layer.
The equipment available for the separation of materials is not equipped with the necessary means of automation of the process, including elements of adaptive management with the adjustment modes according to specified patterns[3].
4. Economic efficiency and effectiveness
Analysis of the known processes and equipment shows that it is possible to achieve a high precision when dividing by establishing regularities of the process under variable processing conditions, the creation of automated systems of process control with adaptation of the parameters, in particular the feed of the tool – disk management state during cutting, the control and adjustment of the position of the cutting in the groove.
The second effective direction of research to reduce the loss of scarce materials is the use of snap-in retention of details until the end of the calibration side surfaces of the groove.
Reveals the prospective use of the results of work in other industries that use precious and scarce materials (medical equipment, dentistry, controls, devices, electrical connectors, etc.), where savings from the elimination of losses of metals can constitute a significant amount.

References

[1] Khafizov I I, Sadykov Z B 2007 Management of process of the combined (electrodiamond) processing of highly rigid materials *Vesntik Kazan state technical university named after A N Tupolev* 1 18-20

[2] Khafizov I I, Sadykov Z B 2012 The analysis of ways of low-waste division of scarce materials with imposing of electric field *Vesntik Kazan state technical university named after A N Tupolev* 2 97-100

[3] I A Konahina, I R Gil'manshin, T R Safin 2014 Heightening of efficiency fuel using installations due toexternal recuperation units of waste heat *IOP Conference Series: Materials Science and Engineering*. **69** 012009

[4] I R Gil'manshin, I A Konahina, N F Kashapov, N N Fahreev 2014 Mini-Central heating and Power Plant (CHP): the choice of the optimal structure and modes of operation *IOP Conference Series: Materials Science and Engineering* **69** 012008

[5] I R Gil'manshin, N F Kashapov 2014 Energy service contracts in regional engineering center for small and medium businesses *IOP Conference Series: Materials Science and Engineering* **69** 012010

[6] N F Kashapov, I R Gil'manshin, I A Konahina 2014 System analysis of the energy complex of engineering enterprise as a basic tool of effective energy management *IOP Conference Series: Materials Science and Engineering* **69** 012024

[7] I A Konahina, N F Kashapov, I R Gil'manshin, R R Ganiev 2014 Optimization of industrial steam supply and steam-and-condensate farming of machine building enterprise *IOP Conference Series: Materials Science and Engineering* **69** 012025

[8] I I Khafizov 2015 Processing methods with imposing of electric field at low- waste division of materials IOP Conf. Series: Materials Science and Engineering **86** 012013

Optimization of energy saving decisions for the large compressor stations of machine-building firms based on the technical and economic indicators

G R Khamidullina

Kazan (Volga region) Federal University
18, ul. Kremlyovskaya, 420008, Kazan, Russia

Abstract
We offer the method of energy-efficiency measures improvement based on optimization of its technical and economic indicators. As an object of research is considered the high-capacity compressor station of the machine-building enterprise.

1.Introduction

Compressor stations of large machine-building enterprises are at the same time the source of energy resource (compressed air) for the production needs and the large consumers of the electric power or other high-quality fuel and energy resources [1, 2]. This causes the continuous search of energy saving decisions at the compressor stations. Energy saving works usually are the mainly concerned with the increase of a share of useful using of the warmth which is taken to the atmosphere by the system of reverse water supply [3, 4]. It could be the utilization of warmth of the water which is taken away from air heat exchangers, from the cooling system of compressors and also the warmth taken in the cooling system of technical oils. The last way is considered in this research.

Now there is the next approach in the practice of researches, development and deployment of energy saving technical solutions: firstly we make an analysis of power efficiency of modernization variants then for the preferable variant we make the technical and economic calculations basing on famous techniques.

Unfortunately such approach could lead to the miscalculations in decision-making and stating of the projects which will not reach the optimum technical and economic indicators. It is expedient to make the search of optimum variant of structure and operating modes of the advanced technical object on the basis of technical and economic optimization with the consideration of real types of the applied equipment and features of its using [5, 6]. Наиболее показательным примером здесь является методология технико-The most attractive example here is the methodology of the technical and economic analysis of combined heat and power plants which founder is the academician L.A. Melentyev [5] distributed in the synthesis of large sources of heat and power supply. This methodology is used here as a basis for the creation of decisions for the research object (compressor station).

IOP Conf. Series: Materials Science and Engineering **134** (2016) 012015 doi:10.1088/1757-899X/134/1/012015

2.Research object

The research object is submitted on the fig. 1 where we could see the initial scheme of the compressor station (CS) of machine-building enterprise with the oil-flooded screws [3].

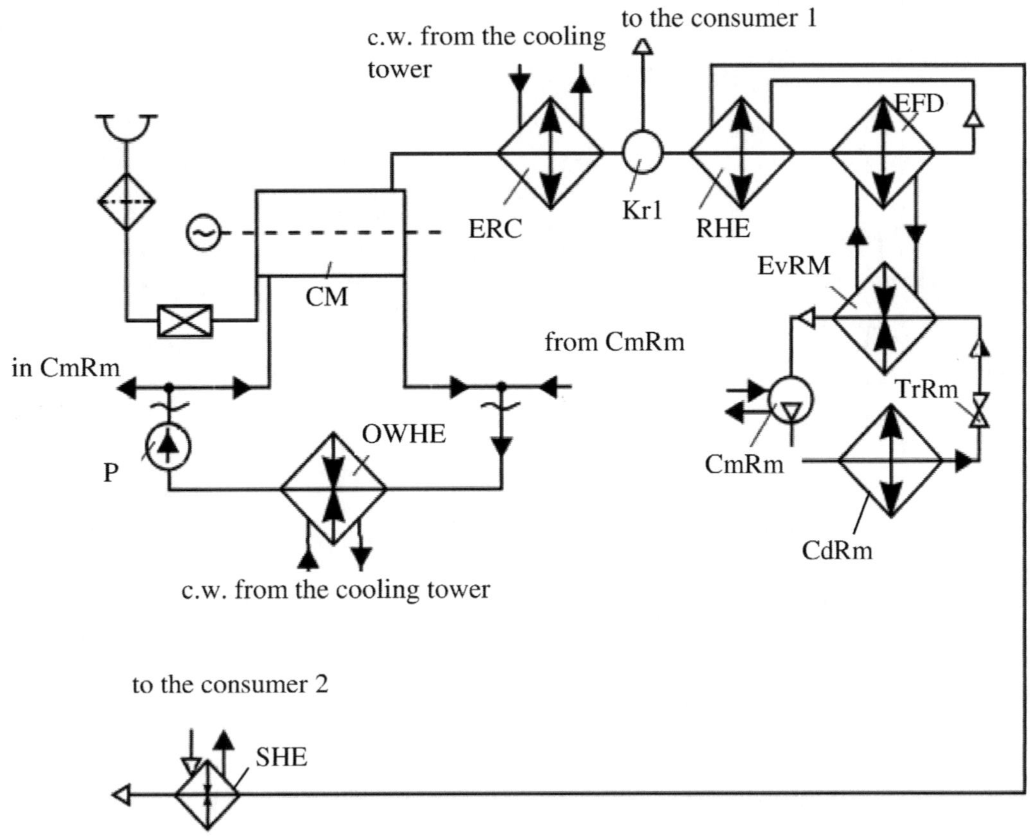

Fig.1. Initial scheme CS of machine-building firm

The scheme includes an air compression site, system of air drainage for ensuring the moist parameters of compressed air demanded in industrial technology, and a site of heating up of the compressed air for the providing the demanded air temperature given to the consumer.

In the CM compressor air compresses up to the pressure demanded to industrial production, and then comes to the system of air drainage. Air drainage is carried out in few steps. The first step – the trailer refrigerator at the exit from the ERC compressor, the second – the RHE regenerative air-air heat exchanger, the third - the EFD heat exchanger-freeze-out device (low-temperature condenser) connected to the RM refrigerating machine.

The cooled and drained air passes through RHE and goes to the heating system. In the initial scheme heating of air up to the demanded on technology temperature of 200°C is carried out in the SHE steam-and-water heat exchanger.

The other symbols accepted in the fig. 1: P - the pump; Kr1 – knot of resetting of compressed air adding; EvRM – the RM evaporator; CdRm – the RM condenser, CmRm – the RM compressor; TrRm – a throttle of RM, OWHE – oil-and-water heat exchanger, c.w. – chilled water.

On the fig. 2 we could see the modified scheme CS [4] where the energy saving solution directed on utilization of warmth of compressor oil in heat pumping installation (HP) with the subsequent using

IOP Conf. Series: Materials Science and Engineering **134** (2016) 012015 doi:10.1088/1757-899X/134/1/012015

of the warmth transformed on higher level for preliminary heating of the air directed then through the heater to the consumer.

In fig. 2 are accepted the next symbols: EvHP – the HP evaporator; KmHP – the HP compressor; TrHP – HP throttle; CdHP – the HP condenser. Other designations are similar to fig. 1.

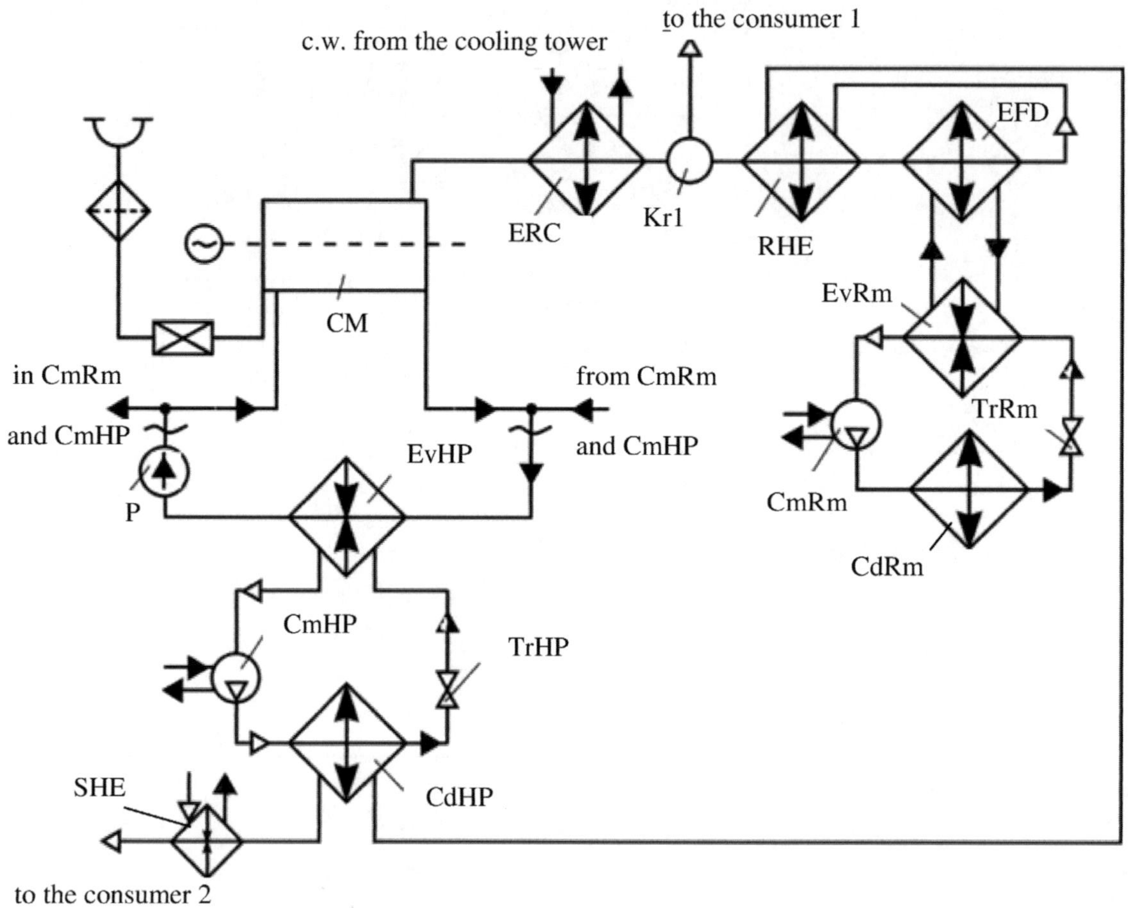

Fig.2. Modified scheme of CS

3.Research methods

The optimum level of intermediate heating in the modified scheme (fig.2.) of air depends on two factors: the positive, connected with direct economy of warmth water vapor, and negative, connected by that in HP additional expenses of the electric power are made for ensuring level of thermotransformation.

Using the analogy with [5, 7, 8] the task of search the most favourable technical or economic parameters combination is formulated as follows:

1. First we are looking for the maximum (minimum) value of efficiency criterion described in the function form:

$$\Phi = \Phi(Z, X)_{\sigma}, \quad (1)$$

with the presence of the nonlinear limiting conditions look like

IOP Conf. Series: Materials Science and Engineering **134** (2016) 012015 doi:10.1088/1757-899X/134/1/012015

$$\varphi_i(Z, X)_\sigma = 0;\ i = \overline{1, n}, \tag{2}$$

and the inequalities like

$$(f_p)_{\min} \leq f(Z, X)_\sigma \leq (f_p)_{\max};\quad p = \overline{1, a} \tag{3}$$

and also with the restrictions on parameters like

$$\left.\begin{array}{l} Z_{\min} \leq Z \leq Z_{\max};\quad Z \in \sigma_m; \\ X \in L_t. \end{array}\right\} \tag{4}$$

Z here is the set of continuously changing parameters (see the p. 3 in the list);

X is the set of discretely changing parameters (see the p. 2 in the list);

σ - set of the characteristics of the considered external factors;

$\varphi = \{\varphi_1, \varphi_2, ..., \varphi_n\}$ - the system of the balance equations for all knots (parts) of system;

$f = \{f_1, f_2, ..., f_a\}$ - the set of technical characteristics on plant knots on which are making the limiting conditions;

the "min" and "max" indexes specify the minimum and maximum values;

σ_m is the dimension of considered space;

L_t is the some final set of discrete elements of *t* dimension.

With the using of identification of consecutive linkages chains between the balance equations setting the values of model parameters there is an opportunity to present the full system of the n-equations in the form of strict sequence of subsystems [5]. Thus it is enough to limit the defining input and output parameters of subsystems (a vector of independent parameters $\overline{X}$) using the conditions like (4), for the reaching of all dependent internal parameters of a subsystem (a vector of Y dimensions of *n*) automatically meet this condition.

Therefore

$$Z = (\overline{X}, Y); \tag{5}$$

provided (that)

$$(\overline{X}_j)_{\min} \leq \overline{X}_j \leq (\overline{X}_j)_{\max};\ j = \overline{1, s};\ s = m - n. \tag{6}$$

One additional feature of the considered systems is that there could be technically the restrictions on nonlinear characteristics on which the criterion function Φ doesn't depend obviously.

As a result the general sequence of task solution is as follows:

1) We define the parameters X and $\overline{X}$, providing the maximum (or minimum) value of efficiency criterion;

2) Then we define the vector Y, proceeding from the received value $X and\ \overline{X}$, as a result of calculation of n-system of equations.

In general terms the equation for definition looks like

$$Y = Y(X, \overline{X}) \tag{7}$$

IOP Conf. Series: Materials Science and Engineering **134** (2016) 012015 doi:10.1088/1757-899X/134/1/012015

3) Functions Φ and f are also defined on each step as the functions of independent parameters X and $\overline{X}$.

With the using of the arguments above the objective is reduced to a problem of finding the function extreme (a minimum or a maximum depending on chosen criterion of technical and economic efficiency).

$$\Phi(X, \overline{X}) \text{ with } (X, \overline{X}) \in R, \tag{8}$$

where R is the area of admissible values set by conditions

$$(f_p)_{\min} \leq f_p(X, \overline{X}) \leq (f_p)_{\max},\ p = \overline{1, a}; \tag{9}$$

$$\overline{X}_{\min} \leq \overline{X} \leq \overline{X}_{\max},\ \overline{X} \in \sigma_s; \tag{10}$$

$$X \in L_t \tag{11}$$

Thus it is supposed that with the fixed values X, functions $\Phi(\overline{X})$ and $f_p(\overline{X})$ are differentiated.

On the analogy with [5] the solution of a task breaks into two stages coordinating the object parameters with the iterative procedures:

Stage 1. Finding the set of the most favourable parameters $\overline{X}$, i.e. the parameters which in the definition range are changing continuously.

Stage 2. The determination of set of the most favourable parameters X, i.e. the parameters which are changing discretely.

4.Results and conclusions

The imposed restrictions look as follows:

1) The potential of the thermal energy developed in HP which is released from HP is limited by the conditions (from conditions of technical feasibility of the decision):

temperature $t''_{c.w.} \leq t^{HP}_{Cd} \leq 125$ °C,

where the $t''_{c.w.}$ is the temperature of the cooled compressed air on leaving the drainage system, °C.

For this parameter is used the designation $\overline{X}_{1i}$, where $i = \overline{1, n}$ is the number of variant of the current settlement combination of these parameters; n – total number of the considered variants.

The second parameter is the temperature level of compressed air at the exit from RHE, $t'_{c.w.}$ arriving then to the drainage system. For this parameter in the model description is used the designation $\overline{X}_{2i}$.

With the parameters $\overline{X}_{1i}$, $\overline{X}_{2i}$ is depend the following chain of the parameters defining technical and economic indicators of the decision:

X_{3i} is the thermal equivalent of the specific consumed electric power of HP, kW;

X_{4i} is an expense of the working agent in HP, kg/s;

X_{5i} is an expense of the heating steam in the SHE heat exchanger, kg/s

X_{6i} is an expense of the refrigerating agent (element) in RM, kg/s

X_{7i} - the thermal equivalent of the specific consumed electric power of HP, kW;

IOP Conf. Series: Materials Science and Engineering **134** (2016) 012015 doi:10.1088/1757-899X/134/1/012015

X_{8i} is an expense of the cooling water in the RM condenser, kg/s

X_{9i} - the temperature of the warmed-up compressed air at the exit from RHE, °C.

As the criterion function was considered the annual economy of the given costs on KS defined by a ratio, roubles/year:

$$\Phi_{ec} = \sum_{i=1}^{n} E_i C_i - \sum_{j=1}^{m} E_j C_j - (K_{mod.CS} - K_{CS})[(\mathrm{E}_i + f) + Z_{add}], \tag{12}$$

where $\mathrm{E}(X, \overline{X})$ is the amount of energy resource; C – energy resource cost; K – level of capital investments; E_i - the accepted coefficient of investment assignments; f – share of annual costs for depreciation, repair of the equipment, salary of workers and other expenses; Z_{add} are the additional unaccounted annual expenses in the modified CS scheme; the $i = \overline{(1, n)}$ - index corresponds to types of the saved energy resources; index $j = \overline{(1, m)}$ – corresponds to types of the spent too much energy resources.

The saved resources in the modified system of CS are:

- thermal energy of steam, t/year;
- reverse water, t/year;
- the electric power spent in system of reverse water supply for pumps and fans of coolers.

The too much spent energy resource in the system is the electric power which is in addition spent for HP drive gear.

The economy or the excessive consumption of energy resources were determined by the balance equations depending on the combination of the above-stated parameters $\overline{X}_{1i}$, $\overline{X}_{2i}$ and $X_{4i} \div X_{9i}$ set in the calculations.

Optimum value of the Φ_{ec} function is seeking for from the condition:

$$\frac{d\Phi_{ec}}{d\alpha} = 0, \tag{13}$$

where $\alpha = [0, 1]$ is a share of the forced-out thermal loading of a steam preheater due to use of the warmth which is taken away in the HP condenser.

The carried-out optimizing technical and economic calculations of the example task showed 20% deviation of the optimum regime parameters in the modified CS which are calculated on thermodynamic indicators towards the reduction of a share of the forced-out α steam [4].

The deviation is concerned with the possibility of accounting the additional factors in the offered method of the technical and economic concerned with the real opportunities of power regulation of HP depending on the thermotransformation mode.

Nevertheless the undiscounted payback period [6] is a little more than one year which confirmed the high efficiency of the proposed energy saving actions.

References

[1] Israfilov Z K and Kashapov N F 1991 *Journal of Engineering Physics* **60** pp. 364–8

[2] Galyautdinov R T, Kasparov N F and Luchkin G S 2002 *Inzhenerno-Fizicheskii Zhurnal* **75** pp. 170–3

[3] Kumirov B A 2006 *Systems of firm supply with the compressed air* (Kazan: Kazan state energy university)

[4] Konakhina I A and Voropayev A N 2009 *Synthesis of power effective system of compressor station with the thermosiphon cooling system of technical oil* (Vologda: Vologda State Technical University) pp. 193-6

[5] Konakhina I A and Nazmeyev Yu G 2002 *Heat-Power Systems and Power balances of Industrial Facilities* (M.: Publishers of Moscow Power Engineering Institute) p. 407
[6] Tabunshchikov Yu A 2005 *Guide on the estimation of economic efficiency investment into the energy saving measures* (M.: AVOK-PRESS) p.120
[7] Fayrushin I, Kashapov N and Dautov I 2014 *J. Phys.: Conf. Ser.* **567** 012009
[8] Dautov G, Dautov I, Fayrushin I and Kashapov N 2013 *J. Phys.: Conf. Series* **479** 012014

Technical and economic analysis of the decisions efficiency of system of water vapor condensate collecting and reusing organization in the industrial firm practice

G R Khamidullina

Kazan (Volga region) Federal University
18, ul. Kremlyovskaya, 420008, Kazan, Russia

Abstract
We consider the technical and economic aspects of power - and resource-recovery decisions on a steam condensate recycling and effective using of the (residual) afterheat containing in it basing on the specifics of the industrial enterprises. Below there are the results of the technical and economic analysis of the original innovative solution introduced in practice of the organization of system of collecting and recycling using of water vapor condensate in the conditions of high probability of its pollution by hydrocarbon joints and a non-return of condensate to the source. Also we showed here the high efficiency of the proposed solution.

1. Introduction

The industrial firms are the main consumers of a water vapor which is used in thermal and power processes. Thus inevitably there is a problem of water vapor condensate return to a source for what at the firms are created the special systems of collecting condensate, difficult on structure leveling its pressure and temperature [1, 2]. Often condensate is polluted and required the ensuring of control of the impurity containing in it and also cleaning methods using. Sometimes danger of pollution is so high that the firm as a source of steam supply refuses to accept condensate for fears of damage of own equipment. In the last case the firm which is a consumer of steam is compelled to assume all the expenses concerned with the collecting and cooling of condensate up to the temperature which isn't exceeding 40^0C due to of ecological safety rules and also its offtake to the reservoirs or special capacities.

With special sharpness this problem is felt at the firms of machine-building and petrochemical branches [3, 4] for which the large volumes of steam consumption are the characteristic – from several hundred to several thousand tons per hour, and the probability of its pollution in use of processing equipment is high and serious.

It is possible to lower the expenses of the firm in such cases through the organization of system of condensate reuse and utilization of the residual warmth containing in it [2, 5].

In spite of the fact that the reached effect of power - and resource-saving within the considered tasks could be shown basing of indicators of power efficiency and non-waste [5, 6] the decision on real measures implementation is made only on the basis of the comprehensive analysis of technical and economic indicators.

Due to carrying out the technical and economic analysis in spite of the fact that there are recommended techniques [7] nevertheless, inevitably there is a need of the considering the specifics of the firm concerned with its specific features, structure, extent of communications, using of innovative engineering solutions, need in account the climatic conditions of the enterprise functioning and the set of other factors [8, 9].

Below there are the results of the technical and economic analysis of the original innovative solution introduced in practice of the organization of system of collecting and recycling using of water vapor condensate in the conditions of high probability of its pollution by hydrocarbon joints and a non-return of condensate to the source [6].

2. The research object

Let's analyze the firm from the petrochemical industry with the high level of industrial steam consumption. The system of condensate collecting and discharge is centralized. In the nominal mode to the condensate point of collecting condensate disposal is about 348,8 t/h of condensate with the average temperature of 950^{0}C and with a residual pressure of 0,25-0,3 MPa.

Due to high probability of condensate pollution by the hydrocarbons its return to a source (CHP plant) isn't made, t.e. the non-return of condensate is about 100%. All formed water vapor condensate is cooled up to the temperature not higher than 40^{0}C then it is partially used for the production needs (about 250 t/h) and all the remained quantity is directed to the sewerage (98,8 t/h).

In the short term on the firm is proposed the start of new productions line that will lead to the serious growth of using the condensate collecting system. After commissioning of new productions the maximum loading of condensate collecting system will increase to 600,3 t/h at the average temperature of the returned condensate at about 97^{0}C.

3. Research data and materials

As the main directions of useful using the condensate warmth for the firm own needs we considered the following actions:

1. The organization of utilization heating of network water in thermal points.
2. The organization of utilization heating of the heat carrier which is released in the hot water supply mode.
3. Production of chilled water in the mode +7 … +12^{0}C on the absorbing bromic and lithium refrigerator (ABLR) during the summer period.
4. Measuring out the heated and cooled condensate based on the requirements of demineralized, softened, partially desalinated and the deep desalinated water for the technological firm needs.
5. As the closing decision on a condensate aftercooling for the reliable work of condenser tank (standpipe) of demanded temperature on 40^{0}C we propose the organization of condensate cooling of in the air-cooled heat exchangers or fin-fan coolers (AC).

Due to marked above actions for an assessment of prospects of the condensate reuse for the technological needs of the firm we made the analysis of possible consumers of warmth, chilled water, partially and deeply desalinated water (PDW and DDW) and also the softened water (tempered water) and relevant requirements of production schedules.

As the most successful decision could be reached only at basing on the combination of above-mentioned actions the search of their best combination assumed the comparative analysis of technical and economic efficiency indicators with the taking into account unevenness and seasonality of consumer's thermal loads.

As the main directions of useful using the condensate warmth for the firm own needs we considered the following actions:

1. The organization of utilization heating of network water in thermal points
2. The organization of utilization heating of the heat carrier which is released in the hot water supply mode;

3. Production of chilled water in the mode +7 … +12^{0}C on the absorbing bromic and lithium refrigerator (ABLR) during the summer period.

4. Measuring out the heated and cooled condensate based on the requirements of demineralized, softened, and partially desalinated and the deep desalinated water for the technological firm needs.

5. As the closing decision on a condensate after cooling for the reliable work of condenser tank (standpipe) of demanded temperature on 40^{0}C we propose the organization of condensate cooling of in the air-cooled heat exchangers or fin-fan coolers (AC).

Due to marked above actions for an assessment of prospects of the condensate reuse for the technological needs of the firm we made the analysis of possible consumers of warmth, chilled water, partially and deeply desalinated water (PDW and DDW) and also the softened water (tempered water) and relevant requirements of production schedules.

As the most successful decision could be reached only at basing on the combination of above-mentioned actions the search of their best combination assumed the comparative analysis of technical and economic efficiency indicators with the taking into account unevenness and seasonality of consumer's thermal loads.

Thus resultant economic effect Eres from the technical solutions realization is determined by the formula:

$$\Sigma E_{res} = E - S, \quad (1)$$

where E – effective (profitable) part of a technical solution, one thousand roubles per year, S – additional operational expenses of a technical solution, one thousand roubles. per year.

With the taking into account income tax Npr the pure firm revenue is determined by a formula, (one thousand roubles per year):

$$\Sigma E_{res} = \Sigma E_{res} (1 - N_{pr}), \quad (2)$$

$$T_{PP} = K / \Sigma E_{res}. \quad (3)$$

3.1. Structure of the modernized system of condensate collecting

In the fig. 1-4 we could see the basic circuit decisions on the organization of system of steam condensate collecting and reusing at the enterprise of petrochemical industry with the indication of parameters of heat carriers.

As it is noted above the using of this system in a year is complicated by the seasonal parameters fluctuations therefore we proposed two main characteristic modes of its work:

1) the heating season (cold season);
2) the summer season.

3.3.1. The work of steam condensate collecting and reusing system during the heating period.

In thermal points I and II (fig. 1) is proposed the utilization heating of heating and heat tracing water due to using of condensate warmth from the productions.

Temperature of the steam condensate returned from productions is rather high (133^{0}C) that allows organizing the step using of warmth with such potential. After the heat exchanger installed in thermal point II condensate with a residual temperature of 96C goes to knot of hot water supply where it is cooled to 15^{0}C, warming up the water for the needs of hot water consumers supply. The decrease of condensate temperature to such level allows using it on feed of the cooler of system of reverse water supply without returning to the point of condensate collecting. Such decision allows us to reach the economy of the electric power spent in pumps for the condensate transportation.

In the collecting condensate point (fig. 2) is made the collecting and quality control of the condensate returned from the water vapor consumers. The calculated temperature of condensate reached in collecting point during the heating period is 85,5^{0}C. The part of condensate of the necessary

IOP Conf. Series: Materials Science and Engineering **134** (2016) 012016 doi:10.1088/1757-899X/134/1/012016

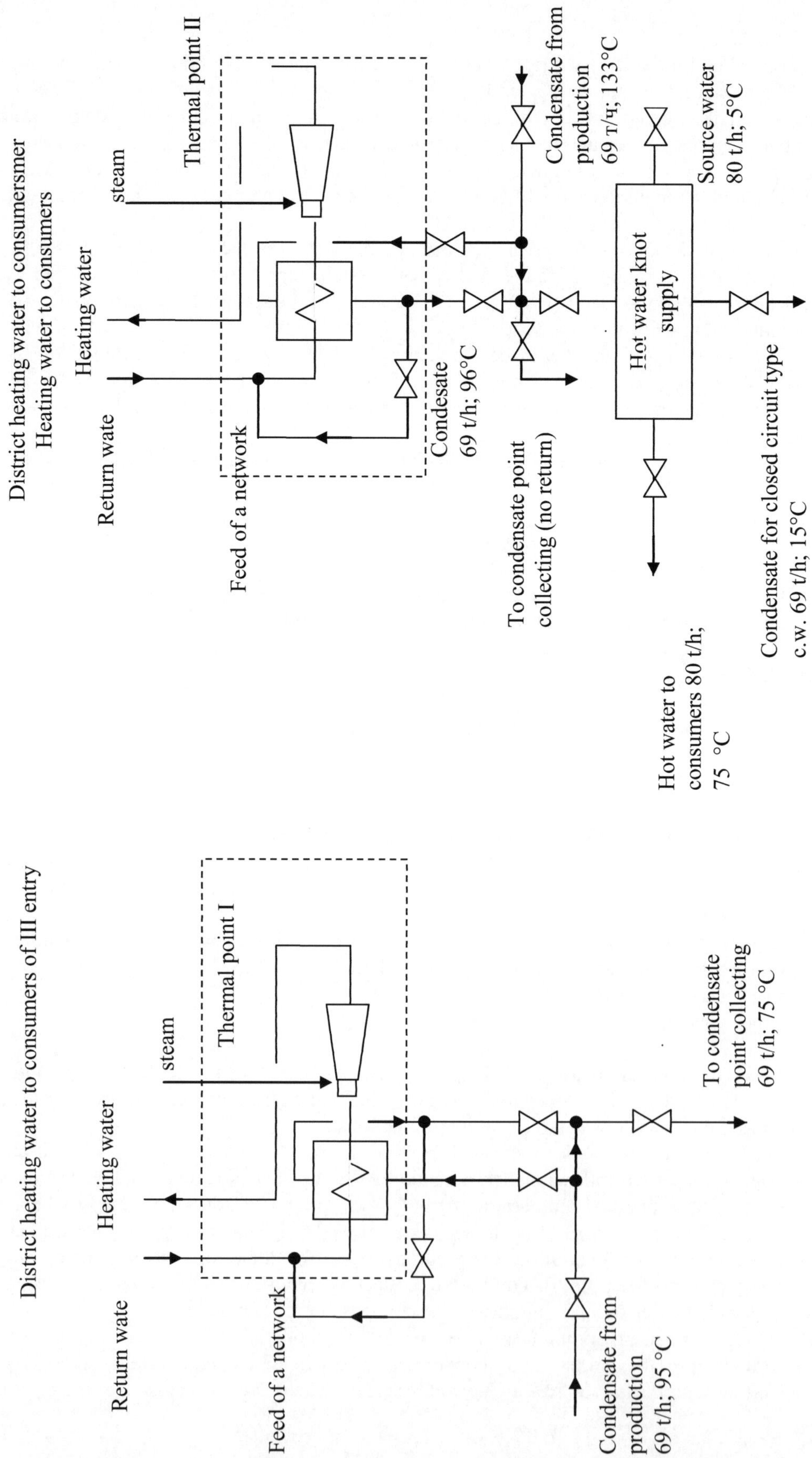

Fig.1. The principal scheme of the condensate warmth utilization in thermal points I and II (heating period)

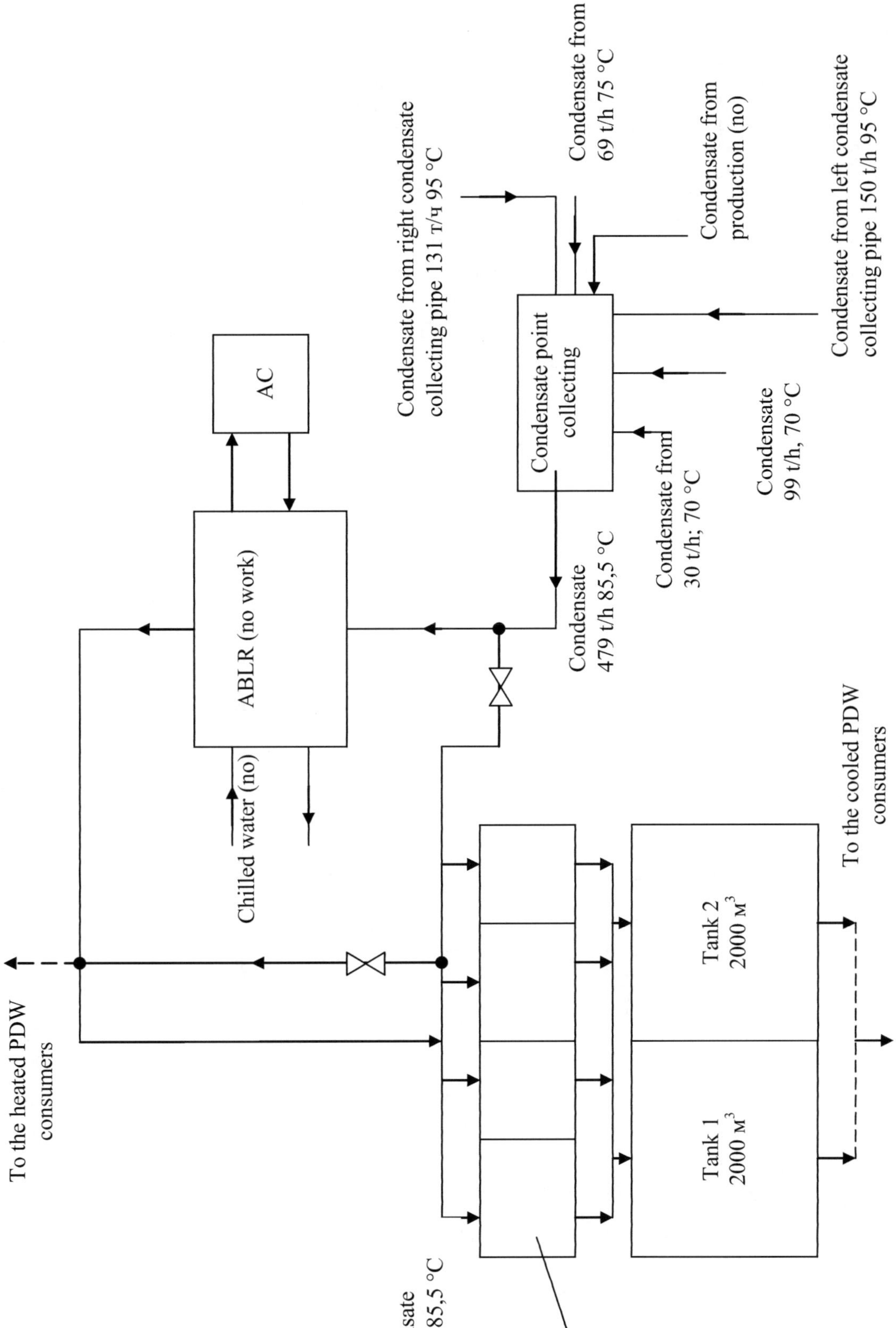

Fig.2. The principal scheme of cooling and reusing of condensate in condensate collecting point condensate (heating period)

IOP Conf. Series: Materials Science and Engineering **134** (2016) 012016 doi:10.1088/1757-899X/134/1/012016

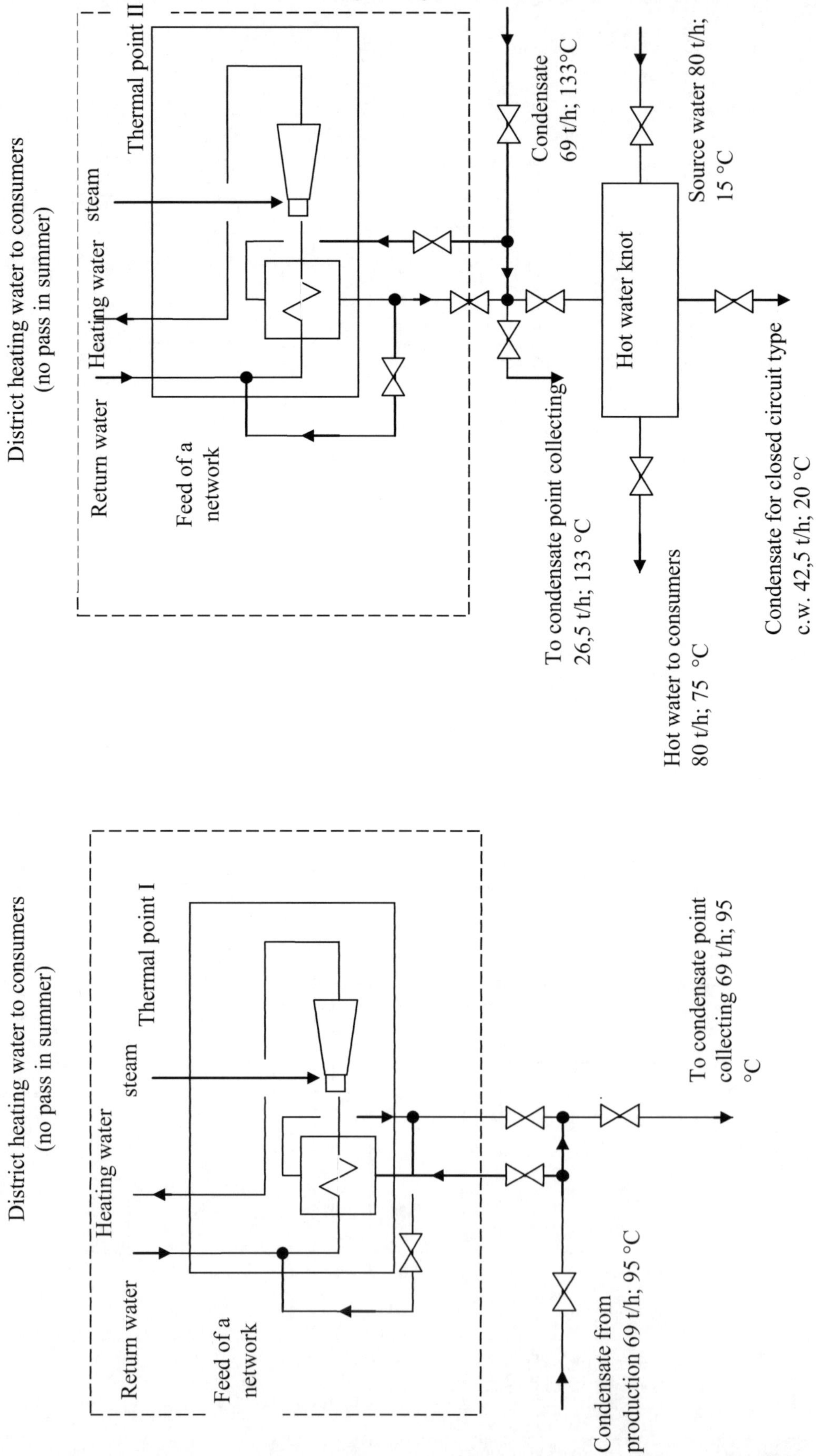

Fig.3. The principal scheme of the condensate warmth utilization in thermal points I and II (the summer period)

IOP Conf. Series: Materials Science and Engineering **134** (2016) 012016 doi:10.1088/1757-899X/134/1/012016

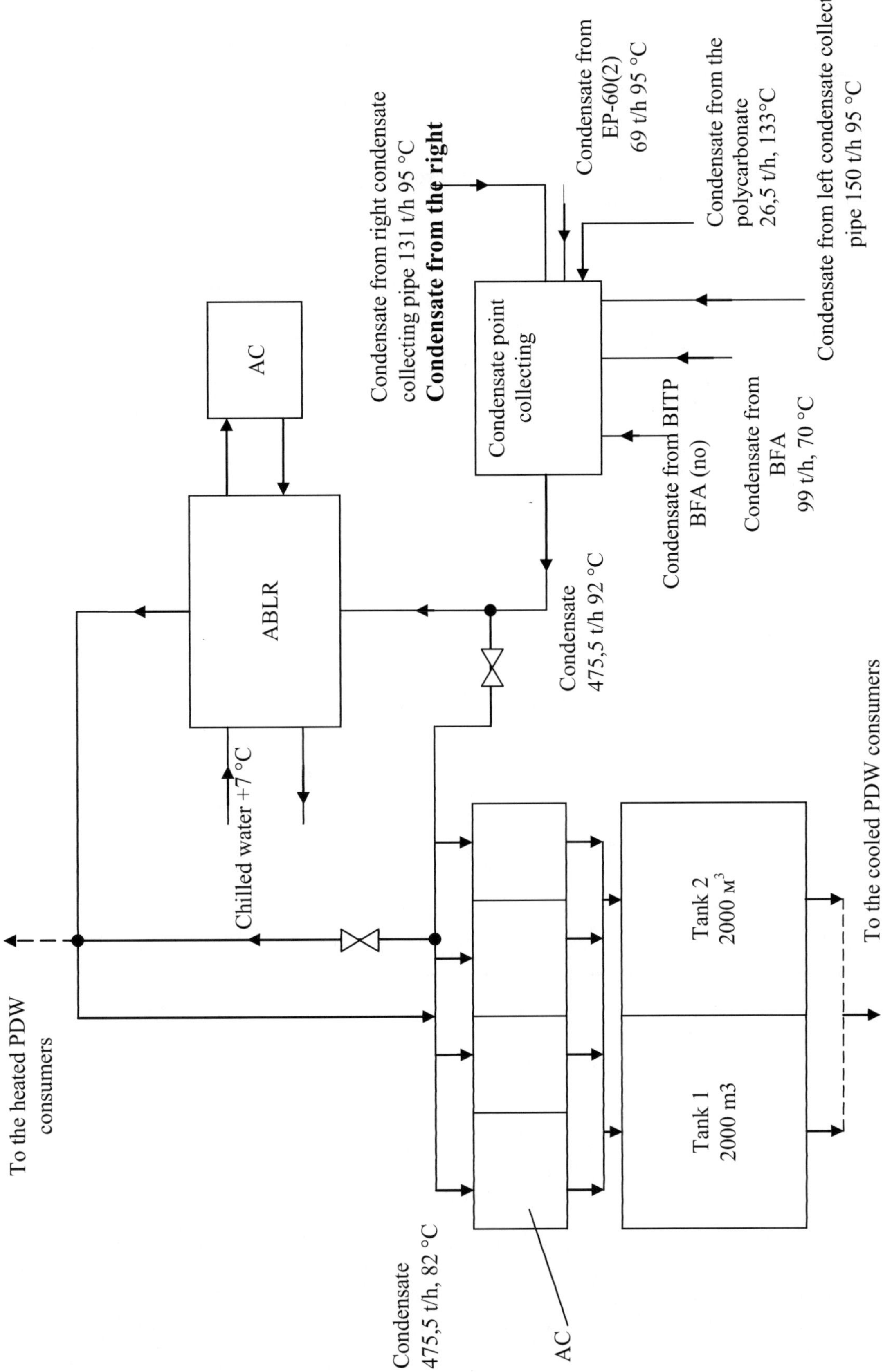

Fig.4. The principal scheme of cooling and reusing of condensate in condensate collecting point condensate (summer period)

IOP Conf. Series: Materials Science and Engineering **134** (2016) 012016 doi:10.1088/1757-899X/134/1/012016

quality could be sent to consumers of heated PDW (about 260 t/h), other quantity is cooled in AC up to the temperature of 30^{0}C.

The cooled condensate melted in capacity from where as required is pumped out to the cooled PDW consumers.

3.3.2. The work of steam condensate collecting and reusing system during the summer period.

During the summer period heating loading on the firm is absent therefore the condensate from productions goes to the collecting point (fig. 4).

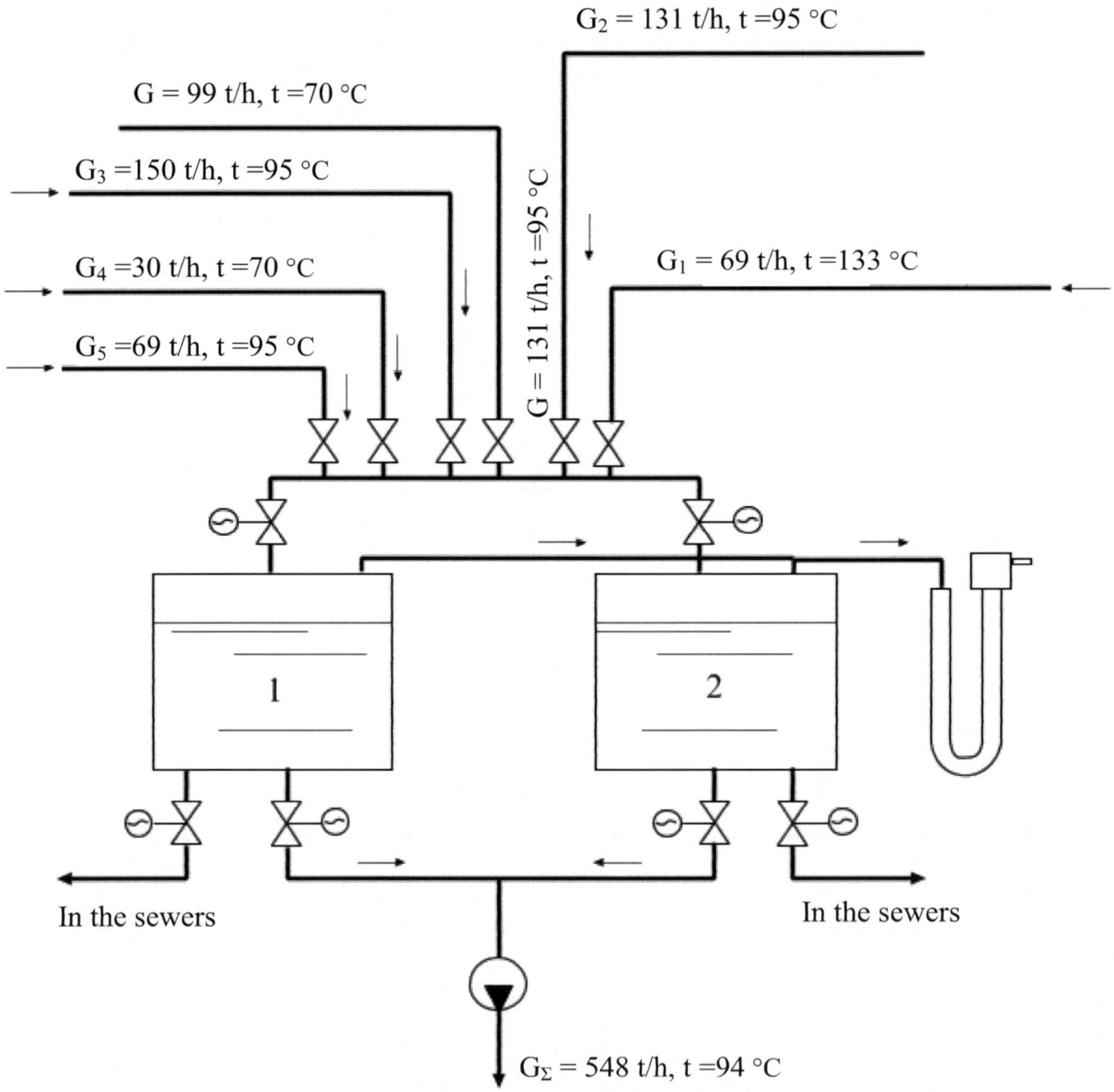

Fig.5. The principal scheme of the organization of condensate collecting point

Loading of hot water supply is year-round, the part of the condensate arriving from the polycarbonate production goes to the knot of hot water supply (thermal point II), and other quantity goes to condensate collecting point.

The calculated temperature of condensate reached in collecting point during the summer period is about 92^{0}C. The part of the warmth containing in condensate is used in the absorbing bromic and lithium refrigerator (ABLR) for the purpose of chilled water production in about 200 t/h for the feed of coolers and an exception of situations of uncooling of return heating water (reverse water) during the maximum temperatures of external air.

At the exit temperature of ABLR condensate is 82^0C. The part of condensate of the necessary quality could be sent to the heated PDW consumers (about 180 t/h); other quantity is cooled in air coolers up to the temperature of 40^0C.

The cooled condensate melted in capacity from where as required is pumped out to the cooled PDW.

The principal scheme of condensate collecting point is given in the fig. 5.

Condensate collecting is carried out to the general collector and for this purpose is proposed to organize the six inserts in a collector for the condensate supply:

- from the right condensate collecting pipe (condensate suction);
- from the left condensate collecting pipe (condensate suction);
- from thermal point I;
- from thermal point II;
- directly from various productions.

For the purpose of an exception of the polluted condensate breakthrough through the collecting system is proposed the installation of two parallel buffer tanks of 1 and 2 of the closed type, each of them is intended for filling time of about 15 minutes ($V = 150$ m^3).

Tanks work in the periodic mode of filling depletion. During the filling of the first tank from the second filled tank, is carried out supply of condensate for the further processing.

Within 15 minutes for which the tank is filled is making in it the selection for the test on the content of organic substances using the express-analysis.

If condensate is sub-standard it is melted in the sewerage opened remotely the necessary tank latch of depletion. Standard condensate with the pump N-1 moves for the further processing to the specialized office.

4. Calculation results

Results of our calculations are presented below in the tab. 1-3. In tab. 1 we could see the capital expenditure for carrying out utilization actions, in tab. 2 – the additional operational expenses of the firm concerned with the implementation of the technical solution. The revenues of the technical solution are given in the tab. 3.

Table 1. Capital expenditures for the equipment.

№	Name of the equipment	Sum of expenses, thousands of rubles
1	Knot of impurity control of in condensate	300
2	Heatexchange equipment	3408
3	Storage containers for thermal energy	244
4	Knot of a chilling of the water vapor condensate	30760
5	Capacities for condensate collecting	500
6	Costs of a construction of additional condensate pipings (lines)	1250
7	Circulators	340
8	Condensate pumps	76
	Total equipment costs:	36878
9	Costs of equipment installation and primary start-up	5532
10	Other unaccounted expenses	3688

	Summary:	46098

Table 2. Additional operational expenses.

№	Name of the equipment	Sum of expenses, thousands rub. per year
1	Depreciation charges	3688
2	Costs of maintenance (re-building)	738
3	Additional costs of the electricity (electric power)	1536
4	Additional salary costs of the enterprise employees	1200
5	Other unaccounted expenses:	1688

Table 3. Revenues from the technical solution.

№	Name of the equipment	Sum of expenses thousands rub. per year
1	The warmth of condensate taken away for needs of hot water supply	18758
2	The warmth of condensate taken away for providing the heating loading of the firm	7675
3	Effect from the replacement with condensate of the softened water	1188
4	Effect from the replacement with condensate of partially desalinated water	58155
5	Effect from the replacement with condensate of demineralized water	18387
6	Additional cold production on ABLR	4525
	Summary:	108688

Basing on the tables 1-3 data resultant economic effect E_{res} from the realization of technical solutions is determined by the formula 4:

$$\sum E_{res} = 99\ 838 \text{ thousands of rub. per year,} \quad (4)$$

with the income tax of 20% the pure firm revenue is (formula 5):

$$\sum E_{res} = \sum E_{res} \cdot (1 - 0{,}2) = 79870 \text{thousands of rub. per year,} \quad (5)$$

with the payback period of capital investments (formula 6):

$$T_{PP} = \frac{46098}{79870} = 0{,}58 \text{ of the year.} \quad (6)$$

5. Conclusions

The payback period of the innovative technical solution on power - and resource-saving in the system of collecting and reusing the water vapor condensate in the conditions of a condensate non-return to a source is less than 0,6 year. As the technical solutions which have a payback period less than three years are identified as an effective [5, 7] the calculated indicator of an innovative solution allows us to make the conclusion about the high economic efficiency of the offered complex of power - and resource-saving actions.

References

[1] Tsvetkov V V 1980 *Organization of steam supply of the industrial firms* (M.: Energy) p. 206

[2] Levin M S 1971 *Use of the fulfilled and secondary steam and condensate* (M.: Energy) p.144

[3] Galyautdinov R T, Kasparov N F and Luchkin G S 2002 *Inzhenerno-Fizicheskii Zhurnal* **75** pp. 170–173

[4] Israfilov Z K and Kashapov N F 1991 *Journal of Engineering Physics* **60** pp. 364–368

[5] Konakhina I A and Nazmeyev Yu G 2002 *Heat-Power Systems and Power balances of Industrial Facilities* (M.: Publishers of Moscow Power Engineering Institute) p 407

[6] Konakhina I A 2011 *Thermodynamic analysis of collecting and condensate reusing system* (IGEU Bulletin) pp. 11-14

[7] Tabunshchikov Yu A 2005 *Guide on the estimation of economic efficiency investment into the energy saving measures* (M.: AVOK-PRESS) p. 120

[8] Dautov G, Dautov I, Fayrushin I and Kashapov N 2013 *J. Phys.: Conf. Series* **479** 012014

[9] Fayrushin I, Kashapov N and Dautov I 2014 *J. Phys.: Conf. Ser.* **567** 012009

IOP Conf. Series: Materials Science and Engineering **134** (2016) 012017 doi:10.1088/1757-899X/134/1/012017

Gas and plasma dynamics of RF discharge jet of low pressure in a vacuum chamber with flat electrodes and inside tube, influence of RF discharge on the steel surface parameters

V I Khristoliubova[1],**N F Kashapov**[2], **M F Shaekhov**[1]
[1] Kazan National Research Technological University, 420015, Russia, Kazan, 68 Karl Marks st.
[2] Kazan Federal University, 420008, Russia, Kazan, 18 Kremlyovskaya st.

E-mail: vallIerrriya@mail.ru

Abstract. Researches results of the characteristics of the RF discharge jet of low pressure and the discharge influence on the surface modification of high speed and structural steels are introduced in the article. Gas dynamics, power and energy parameters of the RF low pressure discharge flow in the discharge chamber and the electrode gap are studied in the presence of the materials. Plasma flow rate, discharge power, the concentration of electrons, the density of RF power, the ion current density, and the energy of the ions bombarding the surface materials are considered for the definition of basic properties crucial for the process of surface modification of materials as they were put in the plasma jet. The influence of the workpiece and effect of products complex configuration on the RF discharge jet of low pressure is defined. The correlation of the input parameters of the plasma unit on the characteristics of the discharge is established.

1. Introduction

Metal cutting tool is widely used by enterprises and ordinary customers. For this reason, very high demands on quality are required, work safety and ergonomics, that in turn forcing manufacturers to create more and more sophisticated types and forms of the instrument.

Metal cutting tools - equipment for different types of work with metal parts. In most cases, these include equipment for various metal-cutting equipment, such as milling cutters, taps, metal knives, saws and drills. Currently, consumers are often faced with a difficult choice to buy an expensive tool under the well-known trademark or cheaper by a little-known manufacturer. In recent years there has been a consistent trend of changing preferences of customers who refuse to buy cheap domestic instrument in favor of more expensive imported. What is the reason? First of all, the fact that the domestic manufacturer is not able to produce a tool that is comparable in quality to foreign analogues, but the lowest price is not so attractive for buyers, because then this advantage is lost because of the need of constant repair, the appearance of injuries and so on.

Manufacturers of cutting tools, feeling the intense competition in the market, introduce more and more new achievements in science and technology to improve productivity, quality and safety of its product. Constructive view of machines and mechanisms is being changed, new alloys for the production are being applied, very actual now is to improve the current cutting properties to improve cross-metal, the possible shortcomings of the current tool are extremely well-calculated to remove them and make it even more perfect.

Milling is one of the most common methods of processing of flat and shaped surfaces. The grooves, ledges, edges, profiles, grooves in the wood products, metal, plastic, acrylic and other materials are carried out with this type of method.

The milling process is characterized by a set of several factors: the rapid rotation of the tool around its axis and the slow translational, rotational or screw sideways movement of feeder. The rectilinear motion of feed is used for the treatment of different types of cylindrical surfaces: planes, all kinds of slots and grooves, shaped cylindrical surfaces. The rotary motion is used for milling surfaces of revolution. Screw traffic flow is handled for screw surfaces such as chip flutes tools, depression helical gears and others.

As the objects of study disk, corner, end mills and drills were chosen.

The technologies currently used in the tool industry and mechanical engineering, have led to the development and wide application of concentrated energy flows such as Plasma Technology. The advantages of these methods are: high density of energy flows; possibility of reactivity and pressure regulation; possibility of ionization of inert and reactive gases for direct modification of the surface layers [1-13].

Analysis of studies on surface modification of metals and alloys shows that the formation of the set of physical and mechanical properties most advantageous to use particles with an energy of 50 - 100 eV. Currents with such energy of ions are formed in jet of radio-frequency (RF) discharges of reduced pressure. In this paper as a high-frequency low-pressure discharge using a capacitive discharge.

Currently, gas and plasmodynamic processes of interaction of plasma flow of RF capacitive low-pressure discharge with metals and their alloys and the effect of reduced pressure RF capacive -setting on the formation of diffusion coatings on the products are not enough studied [14-16]. Properties of RF capacitive jet discharge with flat electrodes are studied very poorly.

2. Results and discussion

For the purpose of steel processing with flow of radio-frequency (RF) discharge of low pressure and the discharge flow studies plasma unit was used. Experienced RF capacitive plasma installation for modification of materials is shown in Fig. 1a.

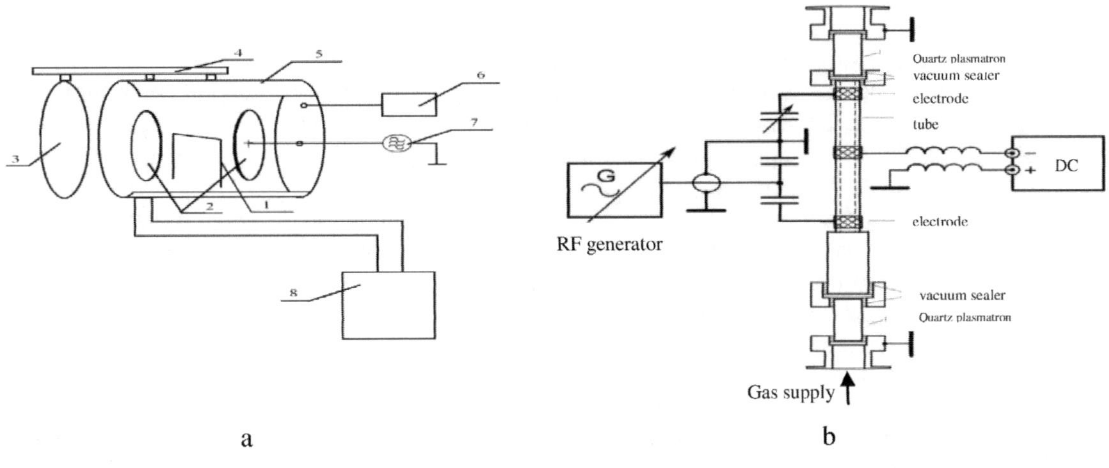

Figure 1. Scheme of the experimental radio-frequency capacitive setup (a), Scheme of tube processing (b). 1 - frame for the installation of the workpiece; 2 - high-frequency electrodes, 3 - the door of the vacuum chamber; 4 - console for opening the door of the vacuum chamber; 5 - vacuum chamber; 6 - supply system and regulation of plasma-forming gas; 7 - RF generator; 8 - vacuum exhausts posts.

RF generator. For studies of different types of RF discharges the creation of RF generators, collected in a single-circuit tuned to the capacitive load on the permitted frequency of 13.56 MHz was required. Consumed power generators range is from 0.5 to 3 kW.

RF plasmotrones. Plasmatron for producing of a plasma flow capacitive type with flat electrodes consists of two water-cooled copper plates. The electrodes are placed in a vacuum unit. Means for fixing samples are located between the electrodes.

Vacuum chamber constructed in a cylindrical design. The door pulls back in a special console with a grounded electrode. Loading is carried into the chamber through the face door in the chamber. When the lid of the vacuum chamber are closed plates are installed in the operating position. Feed through rotation is provided for rotatable products in-chamber volume on the door.

The base of the vacuum unit mounted in a welded frame. Vacuum pumps and water cooling plant components are arranged at the frame unit .

Vacuum exhausts posts consist of two pumps: fore vacuum pump AVZ-20D and twin-rotor pump DVN-150, respectively, with a pumping speed of 50 l / s and 150 l / s. The pressure in the vacuum chamber is controlled by membrane capacitance sensor MKS 627B.

Supply system and regulation of plasma-forming gas consists of a compressed gas cylinder, a reducer for reducing the pressure, manometer, mass flow controller MKS 1179A and a needle valve to regulate flow and a device for producing a mixture of gases.

Water System of Installation BMT-20 serves to provide a set thermal regime of parts and assemblies, the most loaded thermally. The water supply to the installation and removal of it is carried out with rubber hoses.

Control equipment is used to control the input parameters of installation: RF voltage, generator frequency during all experiments for materials processing.

The definition of plasma flow rate, discharge power, the concentration of electrons, the density of RF power, the ion current density, and the energy of the ions bombarding the surface materials and sample processing itself is performed as follows: Samples are hung on the frame (1) and installed between the RF electrodes (2) in a vacuum chamber (5). As the door of the vacuum chamber hanging on the console (4) is closed electrodes are installed in the operating position. Further, the vacuum chamber is pumped using the pumping system (8), built on the basis of a vacuum unit AVR-50. Then, discharge chamber is filled with the plasma gas through the feed system (6) RRG-10. A predetermined gas flow rate is set, voltage is applied from the RF generator (7). The electromagnetic field from the electrodes make a partial ionization of the gas in the chamber, i.e. plasma formation.

Generation of plasma jet inside the tube was carried out according to the scheme shown in Fig. 1b. The tube was set between upper and lower vacuum sealers. The tube is a part of LC-contour and represent the analogue of inductance.

Measurements of plasma in the discharge flow rate of the vacuum chamber with planar electrodes have shown that with increasing gas flow rate increases approximately linearly. The speed of the plasma in the gas mixture on average 5% less than the velocity of the argon plasma. With the increase of the power discharge the flow velocity increases (Fig.2a). A similar dependence occurs with a decrease in pressure in the vacuum chamber. Flow velocity decreases near the metal surface when the article is introduced into the plasma jet. This happens due to the fact that when the body is introduced into the plasma flow in the surface layer flow temperature partially drop due to the difference arising plasma column and the sample temperature. With the same power input to the discharge temperature of the atoms and ions in the gas mixture is smaller than in an argon plasma. Thus, the kinetic energy of the particles significantly affects the discharge flow rate.

The gas-dynamic characteristics of plasma flows in a tube with a diameter ratio to the length of the tubular article 1: 100 were investigated. Fig. 2b shows the dependence of the plasma stream velocity according to the length of the tube in the gas flow rates range G_{Ar} = 1,25-1,86 mg / sec in argon and a mixture of inert and reactive gases (argon + nitrogen). The flow rate along the length of the product increases, because gas pressure in the pumping zone is significantly lower than the inlet zone (Fig. .2c)

IOP Conf. Series: Materials Science and Engineering **134** (2016) 012017 doi:10.1088/1757-899X/134/1/012017

Electron density studies revealed that that the pressure is increased in the range of 10 to 40 Pa electron concentration in the jet RF discharges of low pressure increases. According to the declining of pressure loss of charged particles is reduced due to their diffusion and increasing of the frequency of ionizing collisions with heavy particles. Analysis of ion energy is shown that 10-60 eV is enough for the surface modification.

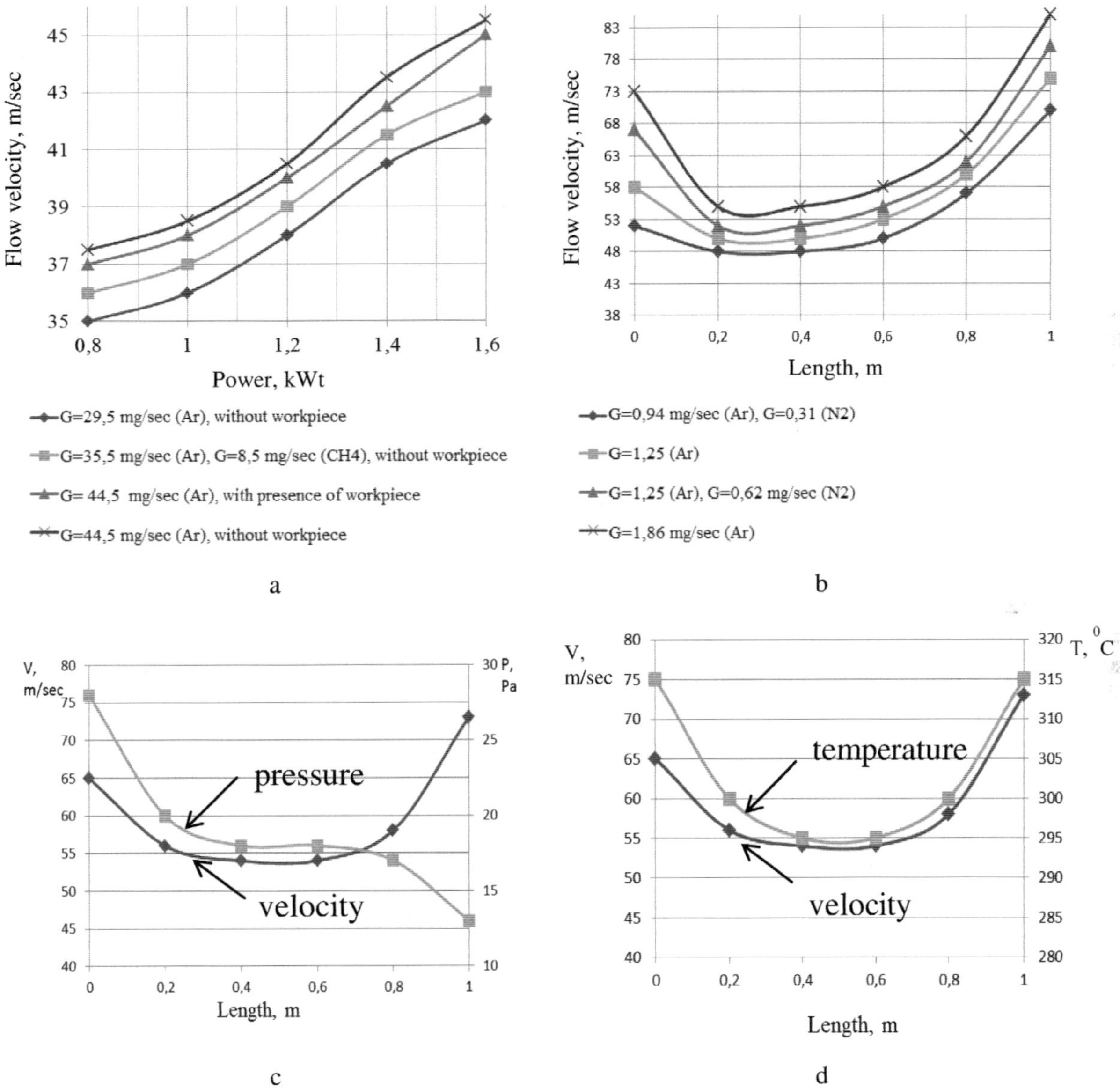

Figure 2. a) dependence of power discharge on the flow velocity in in the range of gas flow rates G_{Ar} = 29,5-41,1 mg / sec in argon and a mixture of inert and reactive gases (argon and methane) in the presence of the workpiece and without the workpiece; b) dependence of gas flow rates according to the length of tube on the flow velocity in argon and a mixture of inert and reactive gases (argon and nitrogen), P=20 Pa, N=1 kWt, c) Correlation of flow velocity and pressure according to the length of tube, 0m – gas supply area, N=1 kWt, G=0,001 g/s, Ar; d) Correlation of flow velocity and temperature according to the length of tube, 0m – gas supply area, P=20 Pa, N=1 kWt, G=0,001 g/s, Ar

IOP Conf. Series: Materials Science and Engineering **134** (2016) 012017 doi:10.1088/1757-899X/134/1/012017

To investigate the effect of the interaction of low pressure RF plasma to the surface of the material experiment with the plasma of an inert gas, argon, and mixture of plasma-chemical gas, methane and argon, was carried out.

Parameters of treatment and results of the experiments are presented in the table 1 and table 2 respectively.

Table 1. Parameters of treatment

N, Watt	Q1, cm^3/min	Q1, mg/sec	Q2, cm^3/min	Q2, mg/sec	P, Pa	Time, min
900	1500	44,4	-		22	20
900	1500	44,4	150	1,8	21	20

Table 2. Results of the experiments

Workpiece (material)	Hardness of untreated work pieces, GPa	Hardness of treated work pieces, GPa
side milling cutter (R18)	13,19±1,93	15,08±0,9
face milling cutter (R18)	14,90±1,02	15,82±1,26
drill (R6M5)	22,5±1,07	30,62±18,53
side milling cutter (R18)	20,87±0,98	20,25±0,87
face milling cutter (R18)	13,28±0,50	23,32±3,97
drill (R6M5)	18,87±0,79	21,42±4,39

Microhardness and roughness measurement was applied to determination of physical mechanical properties. The relief and surface structure at a submicronic and nanometer scale was investigated by means of the scanning nanohardness gage «NanoScan-3D». On the «NanoScan» base the method of measurement of the hardness, based on measurement and the analysis of dependence of loading at indentation of indenter in a material surface from depth of introduction of an indenter is realized. This method is a cornerstone of the standard of measurement of hardness ISO 14577. The indenter of Berkovich type is applied to mechanical tests. It represents a trihedral diamond pyramid with a corner at top near 142°. The method of a measuring dynamic indentation consists in the following: the indenter is pressed into a sample surface with a constant speed, at the achievement of the set loading the indenter is taken away in the opposite direction. In the course of such test record of values of loading and shift of an indenter corresponding to it is made. Feature: pezorezonanse cantilever tuning fork construction with high bending resistance ($\sim 2 \cdot 10^4$ N/m).

For quality control of convex surfaces the device on the basis of nanoskan hardness tester was designed determined specifically for the study of the quality system by the non-destructive methods of physical and mechanical properties of the surface of the pipe.

3. Conclusions

Thus, it revealed that the physical and mechanical characteristics of metal cutting tools treated with capacitive radio-frequency plasma discharge, have a high technological and operational characteristics. Gas saturation (carbonizing) of surface layers of metals and alloys at a depth of 1 micron during processing to 40 minutes was obtained, resulting in an increase of strength properties, durability and lifetime of the products. The advantage of ion implantation over other methods of introducing other impurities in solids is the versatility of the process which allows introducing any element of any material in strictly controlled quantity, as well as setting its depth distribution.

Research of wear resistance was carried out by experimentation way on field trials of "Northwest trunk pipelines." They showed that lifetime of all treated mills increased in the range of 140 to 230%.

The complex approach to the study of surfaces with the use of methods to measure topography, roughness, hardness, wear resistance, modulus of elasticity, elastic recovery coefficient and the thickness of the modified layer in a single instrument was studied and mastered.

References

[1] Dautov G, Dzyba V and Karp N 1984 *Plasma torches with stabilized electric arcs* (Kiev, Naukova Dumka) p 168

[2] Abdullin I Sh, Zheltukhin V S, Sagbiev I R and Shaehov M F 2007 *Modification of nanolayers in Low Pressure Plasma* (Kazan, Publishing House of Kazan. state. tehnol. Univ) p 356

[3] Kashapov N F 2001 *Jet RF plasma torches during the coating process in a dynamic vacuum. Diss. Doctor. tehn. Sciences.* (Kazan) p 506

[4] Sosnin N A, Ermakov S A, Topolyansky P A 2008 *Plasma technology* (St. Petersburg, Publishing House of the St. Petersburg Polytechnic University Press) p 406

[5] Kashapov N F and Sharifullin S N 2015 *IOP Conference Series: Materials Science and Engineering* **86** 012021

[6] Sharifullin S N, Adigamov N R and Slavnina S V 2009 *Bulletin of Samara Scientific Center of the Russian Academy of Sciences. Special. Issue "Actual problems of mechanical engineering"* p 15–18

[7] Dautov G Yu, Kashapov N F, Fayrushin I I and Egorova E A 2014 *Izv. VUZ. Fizika* **57** 97–100

[8] Kashapov N F and Fadeev S A 2014 *Izv. VUZ. Fizika* **57** 116–119

[9] Saifutdinov A I, Fadeev S A, Saifutdinova A A and Kashapov N F 2015 *JETP Lett.* **102** 637–642

[10] Saifutdinov A I, Saifutdinova A A, Kashapov N F and Fadeev S A 2016 *J. Phys.: Conf. Ser.* **669** 012045

[11] Dautov G, Fayrushin I and Kashapov N 2014 *Journal of Physics: Conference Series* **567** 012006

[12] Denisov D G, Kashapov N F and Kashapov R N 2015 *IOP Conference Series: Materials Science and Engineering* **86** 012005

[13] Gaisin R R, Nikiforova A V, Kashapov N F and Kashapov R N 2014 *IOP Conference Series: Materials Science and Engineering* **69** 012007

[14] Simagina E V and Agabeyov Y V 2010 *Proceedings of the Nizhny Novgorod State Technical University named after R.E. Alekseeva* **2** 98–104

[15] Abdullin I, Khubatkhuzin A and Khristoliubova V 2014 *State-of-the-art Trends of Scientific Research of Artificial and Natural Nanoobjects: abstracts for 4th International Scientific conference* (Saint-Petersburg) 161–163

[16] Hubathuzin A A, Abdullin I Sh, Khristolyubova V I and Prokudin S V 2014 *Vestnik of the Kazan Technological University* **17** 39–42

Hybrid Technologies in District Heating Systems

E M Khusnutdinova, I A Konakhina and G R Khamidullina

Kazan (Volga region) Federal University
18, ul. Kremlyovskaya, 420008, Kazan, Russia

Abstract
This paper describes methods of improving district heating systems based on hybrid technologies and presents results of best-performing structure of district heating systems.

1. Introduction

Enhancement of energy efficiency and resource saving are pressing issues worldwide. This issue develops new technologies and critical solutions. Hybrid energy systems, where conventional sources of heat and electrical energy are combined with such non-conventional renewable sources as solar collectors and panels along with the use of alternative fuels, are becoming more common. Such solutions became a best frequent practice, which has a sizeable effect on the economy [1].

Hybrid technologies contribute to the reduction in consumption of conventional fuel and energy resources such as coal, fuel oil, natural gas. External consumption of these resources is reduced thanks to own production of energy and heat that, apart from economic benefits, supports in removing environmental issues, in part or in whole, e.g. reduction in carbon dioxide emissions to the atmosphere [2].

Currently existing hybrid energy systems and their control algorithms are inefficient to meet the requirements of cost effectiveness. They are still behind the conventional systems, particularly, when the prices for natural energy resources drop. However, this solution became popular because these systems have an attractive associated perception. Especially since sophisticated engineering of alternative energy forms is backed at governmental level in the economically developed countries of Europe, Asia and America.

This support resulted in hybrid devices and technologies of their application, which succeeded to combine wide functional requirements. They include hybrid solar collectors, which simultaneously produce electrical energy and heat as they compose a single unit of a solar battery and a solar collector. Unlike the conventional solar batteries, the hybrid devices remove the issue of overheating the photocell. It is known that an overheated photocell stops producing electrical energy [3], while hybrid devices supply the whole excess heat through the heat collector for heating and hot water supply to the buildings. If that is the case, the amount of electrical energy produces highly exceeds the productivity of regular solar batteries.

As compared with the heat collectors, the hybrid solar collector serves longer, produces a low-cost energy unit, requires less operational expenses and has a high thermoelectric performance. In addition, the hybrid collector requires much less area for installation in contrast to the system with separate installation of solar battery array and collectors [4]. This makes such solution suitable for the buildings with limited installation area.

IOP Conf. Series: Materials Science and Engineering **134** (2016) 012018 doi:10.1088/1757-899X/134/1/012018

Another type of hybrid sources is the boiler house that uses both the conventional and alternative fuels (fig. 1) illustrates a typical flow chart of the boiler house.

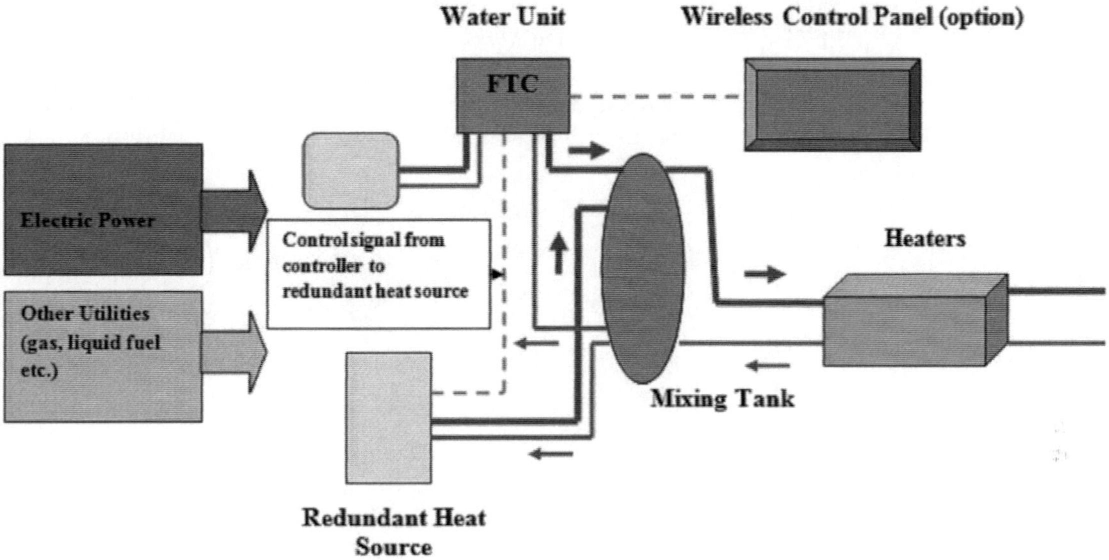

Fig. 1. The hybrid boiler house

The main heat generator of the boiler house is the water-air heat pump, which is so selected that to facilitate the heating requirements of the building or a consumer as well as requirements to hot water supply at ambient temperature ranged -10 to -15°C. If the ambient temperature drops below these values, this will run a fuel or electrical boiler in addition to the heat pump (fig. 2).

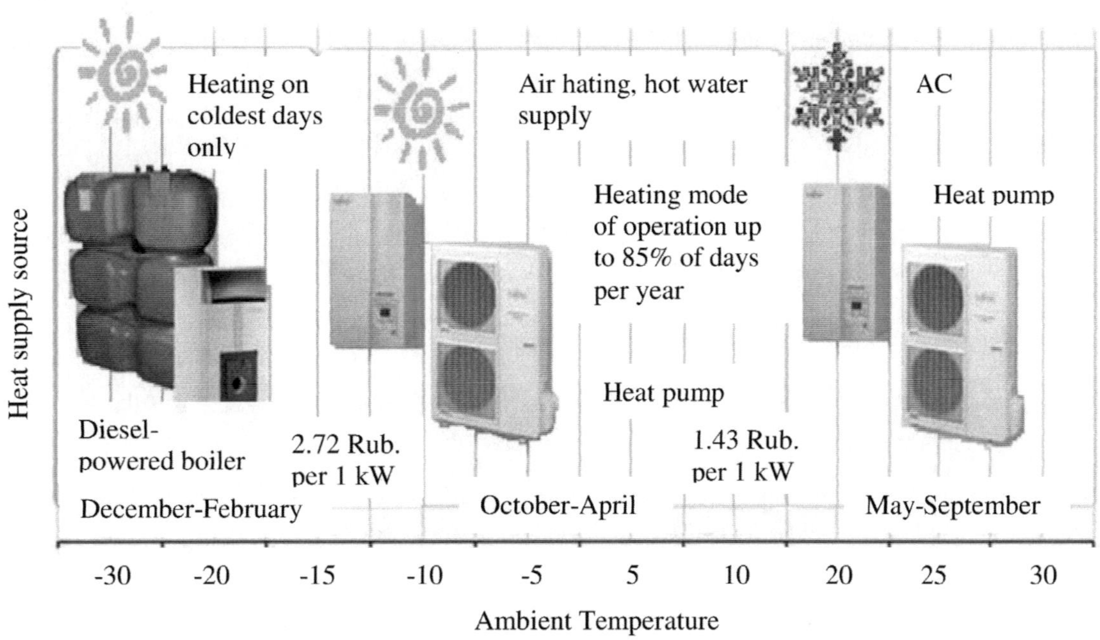

IOP Conf. Series: Materials Science and Engineering **134** (2016) 012018 doi:10.1088/1757-899X/134/1/012018

Fig. 2. A heat pump based hybrid heating system with a diesel-powered boiler

The hybrid boiler houses are cost effective due to inverter heat pumps and reliable as they employ two independent heat generators [4].

This technology is wildly used in individual heating systems, and can be installed in the district heat supply systems.

The comparative analysis of engineering solutions provided a preferred version of hybrid boiler house layout that includes a few options of how to use several alternative sources of energy:

- Water-air heat pump
- Fuel or electric floor- or wall-mounted boiler with weather-dependent automatics
- Fuel burner feed system (if a fuel boiler is used)
- Flue system
- Boiler safety group
- Valved circuit
- Expansion tank
- Indirect-heating hot water tank
- Distribution header with pumps
- Heating circuit makeup
- Hybrid solar collector
- Electric power storage battery
- Secondary equipment

Energy resources can be saved in a district heating system due to the following processes:

- Displacement of external power consumption
- Replacement of thermal load with raw water pre-heating and, partially, with chemically treated water pre-heating
- Displacement of heating load produced by the boiler house and other buildings tied in with the source
- Removal of energy losses associated with the irregularity of installed load due to accumulation of energy resources

3. Conclusions

Optimisation of the hybrid boiler house including more components requires mathematical modelling as well as analysis, synthesis and optimisation of complex heat-power facilities [5].

References

[1] Afonin V S and Pugachyov R V 2014 *Hybrid Power Installations of Multi-Function Fuel Filling Stations* (The Potential of Modern Science)

[2] Mai N G 2013 *Control of Hybrid Power Systems with Renewable Energy Sources* (Author's abstract)

[3] Engineering aids 2015 (*Electronic resource*) URL:http://1haus.ru/info/engineering/

[4] Kadyrov Ch A 2011 *Enhancement of Energy Efficiency of Kyrgyzstan's Heat and Power Systems Using Solar Hybrid Collectors* (Author's abstract)

[5] Konakhina I A and Nazmeyev Yu G 2002 *Heat-Power Systems and Power balances of Industrial Facilities* (M.: Publishers of Moscow Power Engineering Institute) p 407

IOP Conf. Series: Materials Science and Engineering **134** (2016) 012019 doi:10.1088/1757-899X/134/1/012019

Enhancement of Hydrodynamic Processes in Oil Pipelines Considering Rheologically Complex High-Viscosity Oils

I A Konakhina, E M Khusnutdinova, G R Khamidullina and A F Khamidullina

Kazan (Volga region) Federal University
18, ul. Kremlyovskaya, 420008, Kazan, Russia

Abstract
This paper describes a mathematical model of flow-related hydrodynamic processes for rheologically complex high-viscosity bitumen oil and oil-water suspensions and presents methods to improve the design and performance of oil pipelines.

1. Introduction

Production of high-viscosity and paraffinic oils has been steadily increased in the Russian Federation over last two decades, which relates to the deterioration in petroleum supply condition and significant depletion of high-yield fields. Paraffinic oils are produced in promising oil regions in the north of European Russia, Republic of Komi, Yamalo-Nenets and Khanty-Mansi Districts, Western Siberia. Mature fields of Kazakhstan and Turkmenistan are developed in a similar manner.

The process of involving more alike fields into the development is constrained by necessity of searching non-conventional engineering solutions and significant expenditures for production of high-viscosity oils and delivery them to the customers. Primarily, such energy requirements are associated with the need to overcome the viscous friction forces. For that purpose, temperature heating of oils is widely used to reduce their effective viscosity. At the same time, high-viscosity paraffinic oils are rheologically complex fluids frequently exercising pseudo plastic characteristics. This offers challenges to control effective oil viscosity and considerable reduction in friction of piping due to flow share rate.

Intensifiers as regular configurable discrete pipe roughness or inserts different in size are usually used for flow shearing. Such authors as Ye.M. Khabakhpasheva, B.S. Petukhova, G.A. Dreitzer, G. Astarita, G. Marucci, Yu.G. Nazmeyeva, A.Ye. Bergls, O.V. Mitrofanova and others, who study the effects to reduce flow friction in channels with intensifying devices when flowing in viscous as well as pseudo plastic fluids, describe in their papers that vortex intensifiers, e.g. vane, screw, belt vortex generators or those in the form of helical pipes are sufficiently efficient for such flows. Selection of the optimal shape and design parameters of intensifiers depends on many factors – the rheological and thermophysical characteristics of the service environment, climate conditions, extended sections, etc., which requires multivariate calculations and comparative analytical researches [4]. In such circumstances, the processes should be mathematically modelled.

The problem of determining the laws of the hydrodynamic characteristics of abnormally viscous oils flowing in the oil pipelines and active channels is complex and requires systematisation. Although, it is a problem of great engineering importance, there were not enough researches conducted so far, and the present study is aimed to find solution to the identified problem through mathematical modelling. The problem statement is as follows:

We consider the steady laminar isothermal flow of oil in the pipeline with vortex generators [5]. The flow temperature t_0 in the zone of interest is assumed constant; while the averaged flow temperature varies in the range from t_o = 10°C +20°C.

2. The research object

The objectives of the study are set from the standpoint of classical hydrodynamics and with the definition of the components of the velocity vector in a helical coordinate system if the flow integrity, constant oil flow and boundary conditions of flow adhesion on the pipe wall are imposed (velocity $V|_{limit}=0$) [2].

The main problem at the solution of objectives associated with study of a convective heatmass transfer in channels special complexity is hydrodynamic calculation. The solution of the system of hydrodynamic equations includes the equations of movement and continuity makes by the Galerkin's approximation [4].

Let's find the created speed profile at a laminar flow of nonlinear viscoelastic liquid in the channel with a screw knurl. Although the tensor of tension in a timepoint *t* described by nonlinear symmetric tensor functionality from history of deformation and decays in a row Taylor in the vicinity *t=0* on N-orders kinematic tensors of Whyte-Mettsner B_N [6]:

$$B_{N+1} = \frac{d}{dt}(B_N) - (B_N \cdot \nabla V^T) - (\nabla V \cdot B_N), \tag{1}$$

$$B_1 = 2D, \tag{2}$$

where $D = \frac{(\nabla V + \nabla V^T)}{2}$ - the deformation speeds tensor.

Generally view of the set of equation describing this objective and including the equations of movement and continuity:

$$(V \cdot grad = -\frac{1}{\rho} gradP + \frac{1}{\rho} div(T^0), \tag{3}$$

$$div(V) = 0, \tag{4}$$

$$V|_\Gamma = 0, \tag{5}$$

where T^0 – deviator of the stress tensor.

All the borders of taken channel can be provided by family of screw surfaces. A screw channel regardless of cross-section profile possesses one-parametrical group of screw symmetry. As a result required speed field can be considered independent of system of coordinates directed on a channel axis.

To determine the rheological oil characteristics, a differential type model is used, in which the stress tensor relates to the strain rate tensor through a nonlinear function of the effective viscosity dependent on the second invariant of the strain rate tensor.

The stress tensor is defined as:

$$T = -pI + \varphi_1 \cdot (I_2) \cdot B_1, \tag{6}$$

where $I_2 = 4 \cdot tr \cdot D^2 = tr \cdot B_1^2$.

The material magnitude φ_1 can be characterised by sufficiently complex I_2 functions. One of these functions for viscosity ($\varphi_1(I_2)$) is the Kutateladze-Khabakhpasheva model:

$$\varphi_P = \exp(-\tau_P), \tag{7}$$

where $\varphi_P = (\varphi_\infty - \Phi)/(\varphi_\infty - \varphi_0)$,

$$\tau_P = \theta(\tau - \tau_1)/(\varphi_\infty - \varphi_0) \tag{8}$$

where $\tau = \varphi_1(I_2)\sqrt{\frac{I_2}{2}}; \Phi = \frac{1}{\varphi_1 \cdot (I_2)}, \varphi_0, \varphi_\infty - yield.$

To present temperature dependences φ_1, exponential dependence of primary rheological parameters $\varphi_\infty, \varphi_0, \theta, \kappa$ on temperature t is used:

$$\varphi_\infty = A_\infty \cdot \exp(-B_\infty/(Rt)); \tag{9}$$

$$\varphi_0 = A_0 \cdot \exp(-B/(Rt)); \tag{10}$$

$$\theta = \theta_\infty \cdot \exp(-B_\infty/(Rt)); \tag{11}$$

$$\kappa = \kappa_0 \cdot \exp(-C/(Rt)), \tag{12}$$

where $A_\infty, A_0, \theta_0, \kappa_0$ - pre-exponentials;

B_∞, B_0 - activation energy of viscous flow at $\tau \to 0, \tau \to \infty$.

C - constant similar to B_∞, B_0;

R - gas constant.

System of hydrodynamic equations with boundary condition $V|_\Gamma = 0$ has to correspond to an integrated ratio:

$$\iiint_L \{(\rho(\nabla V \cdot V) \cdot \bar{h}) - (div(T^0) \cdot \bar{h}) + (grad\ P \cdot \bar{h})\} dL = 0, \tag{13}$$

where $\bar{h}$ - any element of space of solenoidal vector functions $H_2^1(\Omega)$.

Boundary condition is satisfied for element $\bar{h}$ ($\bar{h}|_\Gamma = 0$). For sequence of basic functions $h^{(1)}$, $h^{(2)}$,.....$h^{(n)}$, belongs to space $H_2^1(\Omega)$, has to correspond:

$$\iint_{\Omega} \{\varphi_1 B_1(V^{(n)}) : D(h^{(k)}) + \varphi_2 B_2(V^{(n)}) : D(h^{(k)}) + + (\rho(V^{(n)} grad\ V^{(n)}) + grad\ P) h^{(k)}\} d\Omega = 0. \tag{14}$$

This system represents the functional operator L_0, defined as $(L_0V, h) = 0$, can be described through the operators L_1, L_2 for which ratios are carried out $(L_1V, h) = 0$, $(L_2V, h) = 0$.

Thus:

$$(L_1V,h) = \iint_{\bar{\Omega}}\{\varphi_1\left[\frac{\partial V^{1(n)}}{\partial q^1}\frac{\partial h^{1(k)}}{\partial q^1} + \left(\frac{\partial V^{2(n)}}{\partial q^1} + \frac{A}{q^{1(2)}}\frac{\partial V^{1(n)}}{\partial q^2}\right)\frac{\partial h^{1(k)}}{\partial q^2} + \left(\frac{2}{q^1}\right)\frac{\partial}{\partial q^2}\left(V^{2(n)} + kV^{3(n)} + \left(\frac{2}{(q^1)^2}\right)V^{1(n)}\right)h^{1(k)} + ((q^1)^2)\frac{\partial}{\partial q^1}\left(V^{2(n)} + kV^{3(n)} + \frac{\partial V^{1(n)}}{\partial q^2}\right)\frac{\partial h^{2(k)}}{\partial q^1} + \left((A^2+1)\frac{\partial V^{2(n)}}{\partial q^2} + Ak\frac{\partial V^{3(n)}}{\partial q^2} + \frac{2}{q^1}V^{1(n)}\right)\frac{\partial h^{2(k)}}{\partial q^2} - 2k^2\varphi_2\right]\left[q^1\left(\frac{\partial V^{3(n)}}{\partial q^1}\right)^2 + \frac{A}{q^1}\left(\frac{\partial V^{3(n)}}{\partial q^2}\right)^2\right]h^{1(k)} + \rho(V^{(n)}grad\,V^{(n)})_1h^{1(k)} + \rho\left(V^{(n)}grad\,V^{(n)}\right)_2h^{2(k)}\}d\Omega = 0, \tag{15}$$

$$(L_1V,h) = \iint_{\bar{\Omega}}\{\varphi_1\left[A\frac{\partial V^{3(n)}}{\partial q^1} + \left(k(q^1)^2\frac{\partial V^{2(n)}}{\partial q^1}\right)\frac{\partial h^{3(k)}}{\partial q^1}\right] + \left[\frac{A^2}{(q^1)^2}\frac{\partial V^{3(n)}}{\partial q^2} + AK\frac{\partial V^{2(n)}}{\partial q^2} + \frac{2k}{q^1}V^{1(n)}\right]\frac{\partial h^{3(k)}}{\partial q^2} + \frac{\partial P}{\partial q^3}h^{3(k)}\}\,d\Omega = 0. \tag{16}$$

Taking into account screw symmetry of a current can say that $\frac{\partial P}{\partial q^3} = C = const.$

$$\frac{\partial P}{\partial q^3} = -\frac{1}{Q}\iint_{\bar{\Omega}}\{\varphi_1\left[\left(A\frac{\partial V^3}{\partial q^1} + k(q^1)^2\frac{\partial V^2}{\partial q^1}\right)\frac{\partial V^3}{\partial q^1} + \left(\frac{A^2}{(q^1)^2}\frac{\partial V^3}{\partial q^2} + Ak\frac{\partial V^2}{\partial q^2} + \frac{2k}{q^1}V_1\right)\frac{\partial V^3}{\partial q^2}\right]\}\,d\Omega. \tag{17}$$

Thus, a differential type rheological state equation has been used for the rheological characterisation of the fluid in constructing mathematical models [1].

A generalised solution to the hydrodynamic equations for the considered conditions can be found in the Galerkin's approximation.

The equations of the Galerkin's approximation are solved using current numerical methods.

Medium-paraffinic oil of the Yuzhno-Khylchuyu field with a mass fraction of solid paraffin 4.61% is considered as service fluid.

3. Conclusions

A twisted tape insert and helical round-shaped wall ledges are taken as intensifying vortex generators. The parameetrs of vortex generators vary between *S/D* = 2-10; *d/D* = 0.6-0.95 (*D* = pipe ID, *d* = reduced diameter of pipe flow area with vortex generator, *S* = twisting step) [3].

Analysis of studying pseudo plastic flow used in the petrochemical industry, can be expected to reduce the effect of the pipe friction for isothermal flows of viscoplastic oils only through a localised impact on the flow at the level of 10-15%.

References

[1] Nazmeyev Yu G 1996 *Hydrodynamics and Heat Exchange Between Vortex Flows of Rheologically Complex Fluids* (M. : Energoatomizdat) p 304

[2] Nazmeyev Yu G 1998 *Heat Transfer at Laminar Flow in Discrete Rough Channels* (M. : Energoatomizdat) p 376

[3] Konakhina I A and Nazmeyev Yu G 1993 *Enhancement of Heat Transfer in the Flow of Viscous Fluid in Helically Rolled Pipes* (*Teploenergetika*) pp 59-62

[4] Konakhina I A and Nazmeyev Yu G *1992 Calculation of Velocity Profile for Nonlinear Viscoelastic Fluids in Helically Rolled Channels* (*Engineering and Physics Journal*) pp 373-79

[5] Vachagina Ye K, Khalitova G R and Nazmeyev Yu G 1965 *On two methods of calculating the velocity profile for non-Newtonian fluids in cylindrical channels with an arbitrary cross-section* (*Engineering and Physics Journal)* pp 65-62

[6] Nazmeyev Yu G 1991 *About equivalence of statements of tasks when modeling a current rheological difficult environments in the auger channels* (*Engineering and Physics Journal)* 2 pp 277-83

IOP Conf. Series: Materials Science and Engineering **134** (2016) 012020 doi:10.1088/1757-899X/134/1/012020

Design of special purpose products made of nanomodified collagen-containing materials with radio-frequency discharge

L Yu Makhotkina[1], **S N Sharifullin**[2]
[1] Kazan National Research Technological University, 420015, Russia, Kazan, Karl Marks st., 68
[2] Kazan Federal University, 420008, Russia, Kazan, Kremlevskaya st., 18

E-mail: lili_makh@mail.ru

Abstract. Research results shows that RF-plasma treatment increases the adhesion of the coating film to the leather uppers and resistance to abrasion and repeated bending of uppers, which define the ability of material to preserve its consumer properties and characterize longer safety of special purpose footwear form during its wearing.

1. Introduction

Textile industry is a significant multiple-discipline sector of the economics, which influences on the stable and balanced social and economic development of the country and provides an independent and intellectual levels of the society.

From the branch work effectiveness point and taking into account of situation around the external economic rate huge and massive tasks connected to the renew of high-quality production and continuous process of labor protection improvement need to be solved by the domestic producers [1,2]. High productivity of labour is not only the result of interaction of human, nature and technique but also right organized labor activity.

Physical activity and capacity for work of the specialist is provided by the rationally chosen equipment. Enterprises provide their workers with means of individual protection, which include garment and footwear of special purpose. Means of individual protection protect from adverse impact during the operation of work. The main aim of use of means of individual protection is prevention of industrial injuries.

Garment and footwear of special purpose are operated in severer industrial conditions than domestic. Because of this the construction and technology of garment and footwear of special purpose have specific features. The level of deterioration of uppers materials are determined by physical activity, which in turn is divided into biomechanical determined by anatomical and physiological features of a man and sanitary and hygienic properties connected to the whole or part hygroscopicity of material [3]. Any means of individual protection, regardless the end use, must be ergonomic, harmless, safe, ensure the unimpeded normal functioning of the human body during the planned period of operation.

Materials used at special purpose footwear production differs a lot from the deterioration of model footwear by character and level of exploitation deterioration and also service life. Character, terms of advent and the intensity of defects development of uppers, lining and bottom of footwear are determined by the sorts and properties of used materials, footwear design, detail disposition and operayion conditions or complex of impacts at the operation.

Russian leather, yuft of chrome style tanning or dense chrome leather of the skins of cattle are usually used for the production of external special purpose footwear details [4]. State All-Union standard 3123-78 "Leather production. Terms and definitions" picks out group of leather, which is considered as leather which is operated in severer conditions. That differs this type of leather from model, which is operated in normal nature conditions without taking into account of professional activity and condition of man. Leathers of special purpose are usually used for the production of clothes and footwear for workers of power structures, energy, oil and gas and other complexes.

Considering the operation and ergonomic characteristics of footwear, the most visible changes of area and volume of uppers take place during its operation. It is explained by the intensive mechanical influence on it. Form and sizes of foot vary during the process of work, so form and sizes of footwear are also vary. By the relying on the beams foot, bending at the phalange joints, expands in this area. At this time foot changes its location and press on the uppers details, which stretch, and then shrink.

Repeated deformations promote the uppers deterioration [5]. In addition in the toe-beam pert of footwear non-vanishing creases are appeared. This leads to the coating surface defects or through-holes on the uppers materials. Besides the stretches and bindings uppers details are exposed to friction, beats, punctures and other impacts.

Product quality and consumer properties of uppers materials depend on adhesion durability of coating to the leather and depend on properties of coating compositions theirselves. Water-soluble polymeric film-forming agents are widely used in methods of protective and decorative finishing of materials surfaces [6]. Leather coating must satisfy the complex of properties, which are defined by the technology of its production and conditions of operation of products made of it. Coating dye, applied to the face surface of leather, penetrates to the elastic porous structure of the papillary layer and must forms the coating which joins the high elasticity with enough durability.

Currently, the existing manufacturing products of special purpose technology of nano-modified leather materials reached their limits. The further development of this direction to create high-quality and competitive products for special purposes can only rely on high technology. These technologies today are plasma technologies. The choice of these technologies should be made taking into account their thermal, cardinality, gas dynamic and energy characteristics [7-18]. Analysis of plasma properties showed that in order to improve the surface characteristics of the materials and increase the top coating adhesiveness and polished natural leathers in this case is the most suitable radio-frequency plasma reduced pressure [19-23].

2. Results and discussion

Polished and original leather was activated by the radio-frequency plasma (RF) of low pressure to increase surface characteristics of uppers materials and adhesion [7,8]. Modification of the coating film adhesion during the surface treatment of polished and original leather by the RF-plasma is shown in fig. 1. Coating adhesion to the leather increased in 3-4 times after the 3 minutes treatment (fig. 1(a)). Coating adhesion to the leather with origin face surface increases as the power of the charge increases up to 1.07 kWatt (fig. 1(b)); to the polished leather as the power of the charge increases up to 1.3 kWatt.

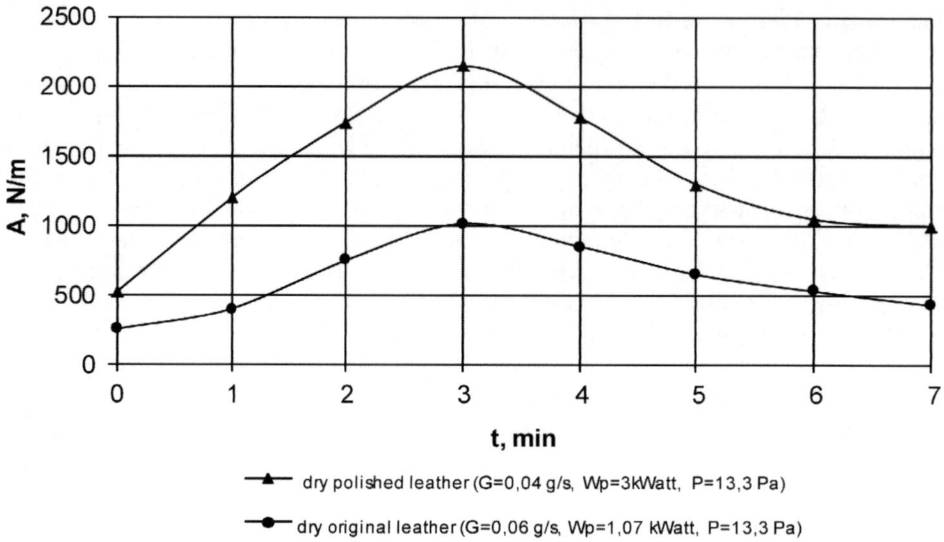

a)

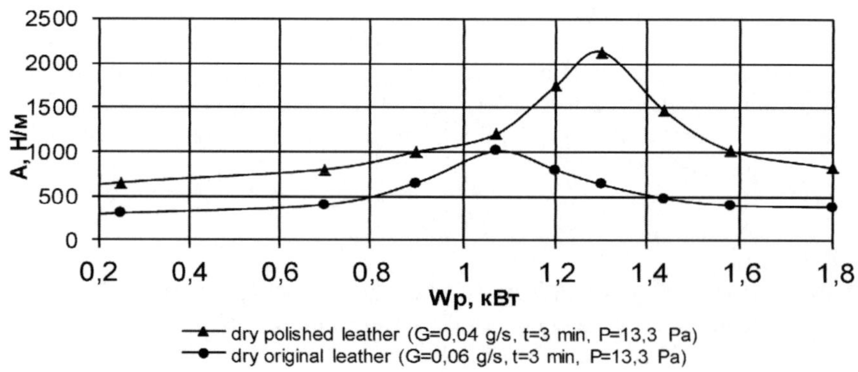

b)

Fig.1. Modification of adhesion to the modified leather depending on: a) time of plasma treatment; б) charge power

To define the influence of the RF-plasma stream to the acryl film the following researches were carried out. A film without being applied to the leather was treated by the plasma.

Table 1. Coating adhesion durability to the original and polished modified leather.

Leather type	Adhesion incensement in times	
	Dry condition	Wet condition
Original leather with emulsion coating	3,7 1,5	2,4 1,3
Polished leather with emulsion coating	4	3,2

Plasma treatment leads not only to the modification of interaction of film and leather but also to the structure modification of the film itself. According to the data of the table 1 adhesion durability of coating film after the RF treatment for original and polished leather increased, especially in a dry condition. At the same time tensile strength was increased by 15-20%.

For the regularities research of plasma influence on the coating dying of leather mechanical indicators of leather were defined as the operational properties are one of the basic properties and characterizes quality of products and their destination. RF plasma treatment leads to the incensement of different mechanical properties of leather after applying of the emulsion coating (table 2).

Table 2. Influence of plasma treatment on the operational characteristics of leather with emulsion coating: a – polished leather, b – original leather.

Indicators	With RF-plasma		Without RF-plasma	
	a	b	a	b
Tensile strength, MPa	17,0	18,4	14,7	15,2
The stress at cracking of an face layer, MPa	16,2	17,7	13,9	14,5
Elongation at tension 10 MPa, %	35	35	36	37
Resistance of the coating to repeated bending, points	4	4	3	3
Coating film adhesion, N/m - to dry leather - to wet leather	 920 421	 2433 887	 230 133	 656 371
Abrasion resistance, turns	60	50	50	40

Tensile strength was increased by 10-20%, the strength of the front layer by 10-20%, resistance of the coating to repeated bending by 35%, abrasion resistance by 20-25% after the RF-plasma treatment.

3. Conclusions

Research results shows that RF-plasma treatment increases the adhesion of the coating film to the leather uppers and resistance to abrasion and repeated bending of uppers, which define the ability of material to preserve its consumer properties and characterize longer safety of special purpose footwear form during its wearing.

References

[1] Nikitina L L, Makhotkina L Yu and Fukina O V 2010 *Leather and foot-wear industry* **(3)** 24–25

[2] Makhotkina L Yu 2012 *Herald of Kazan technological university* **15** 255–258

[3] Makhotkina L.Yu. 2006 *Regulation of forming ability of complex materials of footwear industry using the nonequilibrium low-temperature plasma. The thesis for the degree of Doctor of Technical Sciences* (Kazan) p 341
[4] Makhotkina L Yu and Khristoliubova V I 2015 *Inorganic Materials: Applied Research (Physics and Chemistry of Materials Treatment)* **(4)** 35–38
[5] Sirazijeva L F, Stepin S N and Makhotkina L Yu 2004 *Lakokrasochnye Materialy i Ikh Primenenie* **(10)** 25–28
[6] Osin Yu N, Makhotkina L Yu, Abutalipova L N and Abdullin I Sh 1998 *Vacuum* **51** 221–225
[7] Saifutdinov A I, Saifutdinova A A, Kashapov N F and Fadeev S A 2016 *J. Phys.: Conf. Ser.* **669** 012045
[8] Saifutdinov A I, Fadeev S A, Saifutdinova A A and Kashapov N F 2015 *JETP Lett.* **102** 637–642
[9] Kashapov N F and Sharifullin S N 2015 *IOP Conference Series: Materials Science and Engineering* **86** 012021
[10] Kashapov N F and Sharifullin S N 2015 *IOP Conference Series: Materials Science and Engineering* **86** 012022
[11] Israfilov Z K and Kashapov N F 1991 *Journal of Engineering Physics* **60** 364–368
[12] Dautov G, Fayrushin I and Kashapov N 2014 *J. Phys.: Conf. Ser.* **567** 012006
[13] Luchkin A G and Kashapov N F 2014 *J. Phys.: Conf. Ser.* **567** 012027
[15] Dautov G, Dautov I, Fayrushin I and Kashapov N 2013 *J. Phys.: Conf. Ser.* **479** 012001
[16] Dautov G, Dautov I, Fayrushin I and Kashapov N 2013 *J. Phys.: Conf. Ser.* **479** 012014
[17] Galyautdinov R T and Kashapov N F 2003 *Svarochnoe Proizvodstvo* **(3)** 27–31
[18] Israfilov Z K and Kashapov N F 1987 *Soviet Aeronautics* **30** 97–99
[19] Abdullin, I.Sh., Galyautdinov, R.T., Kashapov, N.F. 2001 *Inzhenerno-Fizicheskii Zhurnal* **74** 104–107
[20] Abdullin I, Voznesensky E, Sharifullin F, Dzhanbekova L 2010 *News of Higher Schools. Technology of light industry* **8** 12–15.
[21] Abdullin I, Krasina I, Shaekhov M, Sharifullin F 2004 *Applied physics* **(6)** 59–65
[22] Sharifullin F S 2011 Scientific-technological basics of fur production with the regulating operational properties using the nonequilibrium low-temperature RF discharge plasma. The thesis for the degree of Doctor of Technical Sciences (Kazan) p 321
[23] Kashapov L N, Kashapov N F and Kashapov R N 2014 *J. Phys.: Conf. Ser.* **567** 012024

Application of wave coagulation of the aerosol in technology of gasification of cryogenic liquids

B R Mubarakshin, V G Tonkonog, S N Arslanova, A L Tukmakov and D A Tukmakov

Kazan National Research Technical University of A. N. Tupolev-KAI, Russia

E-mail tukmakov@imm.knc.ru

Abstract. Process of acoustic coagulation of a polydisperse gas-suspension under the influence of a nonlinear wave field in relation to technology of gasification and cryostatting of the liquefied natural gas and hydrogen is investigated. The gas-suspension of polydisperse structure is formed in the stream proceeding through Laval's nozzle then the bearing and disperse phases are divided in an inertial separator. For increase of efficiency of division of phases between a cut of a nozzle and an entrance of a separator the working site in which on the polydisperse environment acoustic impact is made settles down.

In work [1] was offered power - and resource-saving technology of gasification and cryostatting of the liquefied methane. The technological scheme of installation is given in figure 1. Liquid from the heat tank 1 moves in the heat exchanger 2 where liquid heats up to temperature of T_1, below boiling point temperature and moves in Laval's nozzle where in the vicinity of the minimum section there is its low-temperature boiling up to the subsequent inversion of structure of a stream and formation in the extending part of a nozzle of polydisperse vapor-drop current. Experiments show that temperature, which is below triple-point temperature, might be acquired by variation of nozzle inlet temperature, pressure drop over nozzle and nozzle profile. As a result, in the extending part of a nozzle the disperse phase will consist of liquid drops and firm particles with a temperature, lesser than temperature of liquid in capacity storage. Then vapor - drop stream moves in an inertial separator where the bearing environment - gaseous methane separates from a disperse phase which then downloads the pump 8 in storage and lowers the liquid temperature that reduces intensity of losses and increases a possible period of storage of methane in a liquid state. The bearing phase - vaporous methane goes to the heat exchanger 5 where heats up to temperature of T2 which is required for operation of the power station 7.

Efficiency of process is defined by quality of separation of the bearing and disperse phases. Experimental data show that at the exit of a nozzle there are particles with sizes vary from micron to tens microns. At the same time, inertial separators are capable to separate a disperse phase from the bearing environment if the sizes of particles exceed about 5 microns. In this regard there is a decrease in efficiency of technological process – in capacity the smaller number of the cooled disperse fraction which partially gets to the heat exchanger 5 where on its heating to melting temperature, on melting process, having heated to evaporation temperature downloads, the subsequent evaporation and heating to temperature of T_2 are required additional expenses of energy. Therefore, on a site between day offsection of a nozzle and entrance

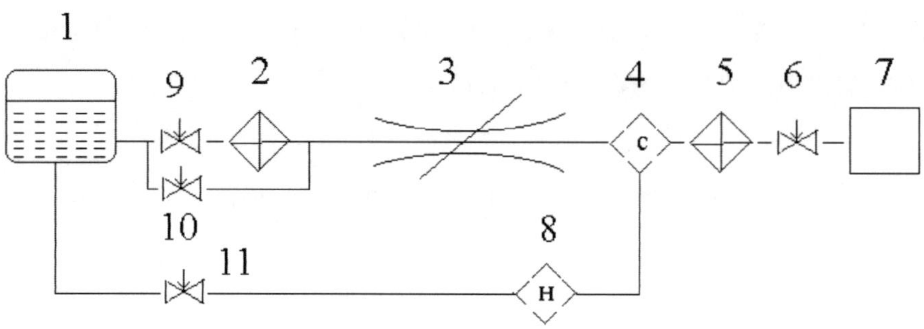

Figure 1. The scheme of installation for gasification and cryostatting of the liquefied methane: 1-heatisolated capacity with the liquefied natural gas; 2- heat exchanger heater of liquid; 3-nozzle of Laval; 4-inertial separator; 5- heat exchanger heater of steam; 6, 9-11-gates; 7-power station; 8- pump.

of an inertial separator (figure 2) it is necessary to change dispersion of vapor - drop stream, having reduced the mass maintenance of fine fractions due to coagulation of particles. One of possible ways of acceleration of coagulation is wave impact on vapor - drop stream [2]. Thus the choice of intensity of wave influence is limited to critical number of Weber [2] in order that coagulation of particles at impact prevailed over their crushing. The model in which the fraction [3, 4] is compared to each size of particles is applied to the description of dynamics of a coagulating polydisperse gas-suspension.

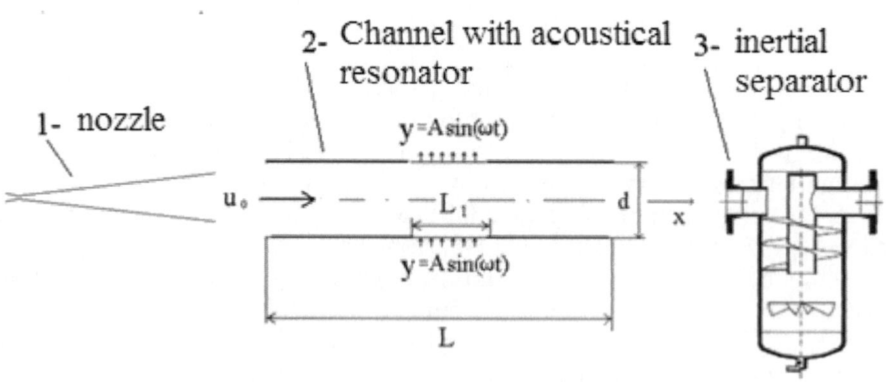

Figure 2. The scheme of an arrangement of a nozzle - the steam generator, the channel with the acoustic resonator and an inertial separator.

The bearing phase is described by system of the equations of Navier-Stokes considering an exchange of an impulse and energy with the disperse phase including n of fractions each of which described by system from the continuity equation for the average density, the equations of preservation which are working out an impulse and the equations of preservation of internal energy [4-6]. Coagulation of particles is described by the Lagrangian model considering an exchange of weight, an impulse and energy between fractions as a result of impact of particles [2, 4]. In an initial time point of t=0 vapor-drop mix from five fractions of drops of liquid methane with radiuses of particles of r_{10}=1mkm, r_{20}=5mkm, r_{30}=10mkm, r_{40}=15mkm, micron r_{50}=20, moves with u0 speed, evenly filling the flat channel (fig. 2). At t>0 process of inphase fluctuations of a cross component of speed of sites of the top and lower walls of the channel $((L-L_1)/2<x<(L+L_1)/2, y=0)$, $((L-L_1)/2<x<(L+L_1)/2, y=d)$ under the law $v(t)=A\cdot sin(\omega t)$(fig. 2),leading to generation of fluctuations in the bearing and disperse phases with the first own frequency of rather cross section of the channel$\omega=\pi c/d$, where $c=(\gamma RT)^{1/2}$–assessment for sound speed in methane vapors at the set temperature.

In figure 3, a, b dependences of cross and longitudinal components of speed of the bearing and disperse phase (5 fractions) on time in a stream which the field of a standing wave with an amplitude of fluctuations of a cross component of speed on A=0,0005m walls locally influences are presented. Radiuses of particles of all fractions, except the smallest, grow eventually (figure 3, c). The average density of fraction with a radius of particles of 1 micron decreases eventually and by the time of t=0,035 c with becomes close to zero. The average density of larger fractions (5, 10, 15 microns) at first reaches the greatest value then decreases. The average density of the largest fraction continuously grows eventually (figure 3, d). The highest speed of reduction of concentration of particles owing to coagulation in an acoustic field at the smallest fraction. In the vicinity of a source of acoustic fluctuations concentration of particles with a radius of 1 micron at t = 0,01 with falls approximately on 4 orders (figure 3, e). Effective radius of particles of all fractions except for the smallest quickly increases in the same area and the zone extended down a stream in which the average density of the fine decreases is formedfractions the average density of fraction of the largest particles from R_0=20 microns also increases (figure 3, e).

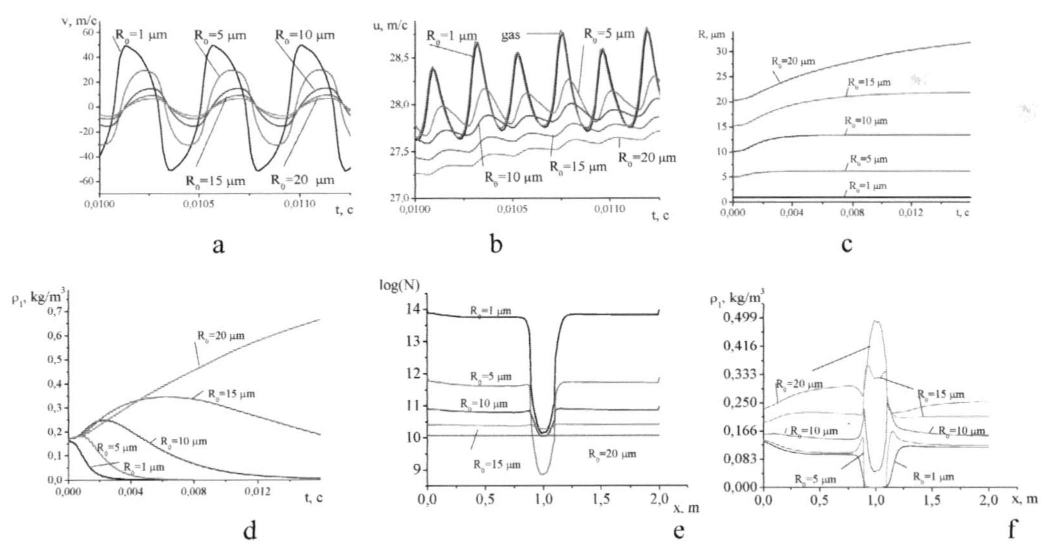

Figure 3. Dependence on time cross (a) and axial (b) the making speed of particles, their effective radius (c) and the average density of fractions (d) in a point (x=L/2, y=d/2), concentration (e) and average density (f) particles of fractions along a channel axis.

Thus, impact on vapor -drop stream of poly-disperse structure resonant wave fields allows to reduce concentration of fine fractions of mix due to coagulation of particles of different fractions. For this reason between a cut of a nozzle generator of a poly-disperse gas-suspension and an entrance flange of an inertial separator expediently placement of the acoustic channel resonator in which there is an integration of particles due to coagulation that increases quality of division of the bearing and disperse phase in inertial separators.

Work is performed with financial support of the Ministry of Education and Science of the Russian Federation, the Agreement on a subsidy No. 14.577.21.0151 of 28.11.2014. The identifying number of the project RFMEFI57714X0151.

References

[1] Tonkonog V G and Arslanova S N Sistema of supply of cryogenic fuel in the power station. *Patent Russian Federation for the invention of Ru.* No. 2347934. Date of issue 10.06.2009.
[2] Alemasov V E and Dregalin A F and Tishin A P and Khudyakov V A 1971 *Termodinamicheskiye and heatphysical properties of products of combustion. The reference book in five volumes.T.1. Calculation methods.* Moscow VINITI production publishing combine
[3] Kutushev A G 2003 *Mathematical modeling of wave processes in aero disperse and powdery environments* St. Petersburg Publishing house Subsoil
[4] Tukmakov A L 2015 *Journal of Engineering Physics and Thermophysics* **88** 9-17
[5] Saifutdinov A I, Saifutdinova A A, Kashapov N F and Fadeev S A 2016 *J. Phys.: Conf. Ser.* **669** 012045
[6] Saifutdinov A I, Fadeev S A, Saifutdinova A A and Kashapov N F 2015 *JETP Lett.* **102** 637–642

IOP Conf. Series: Materials Science and Engineering **134** (2016) 012022 doi:10.1088/1757-899X/134/1/012022

Experimental investigation of highly non-linear fluctuations of gas of an open pipe in the vicinity resonances

L R Shaydullin[1]

[1] Kazan Federal University, 18 Kremlyovskaya str., Kazan 420111, Russia

E-mail: liqn@mail.ru

Abstract. Non-linear oscillations of gas in an open pipe were researched experimentally. Dependencies of the pressure oscillation amplitude were obtained for various frequencies of the gas excitation near the first proper frequency to the approach to the shock-wave oscillation mode.

Introduction

The development of modern technology brings a number of challenges, among which is one of the actual study of the resonant vibrations of the medium in different apparatuses of power engineering, chemical technology. Of particular interest is the systems of a resonator-pipe type, which a widely used in technics, where one end of the pipe communicates with the environment. Highly non-linear pressure waves emerge in pipe systems near the resonance and non-linear effects are observed [1, 2], such as turbulence in the flow, the occurrence of secondary flows and acoustothermal processes, generation of higher harmonics. The results of theoretical and experimental studies of these environments are analyzed in reviews [3, 4], where he studied modes of unstressed wave currents at small amplitudes of excitation of gas [5-7], and with the formation of shock waves with large amplitudes.

In this new series of studies on the based company vibration generator TIRAvib 5220 / LS with digitally processing were considered different gas dynamic characteristics of highly nonlinear oscillations of gas, which have been little studied.

1. Experimental setup

The research was performed in an installation based on a vibration generator 1 of TIRAvib S 5220/LS make with a power amplifier 15 of BAA 1000-ET model by TIRA firm (Figure 1). Into the vibration table 3 was screwed piston pin flat piston 5 and the diameter d = 100 mm; the total weight is 250 g.

The piston enters the cylinder 6 and oscillates according to the harmonic law with the set frequency ν in Hz. In order to prevent the transition of the stand vibration to the floor, the installation was placed on pneumatic spring cushions 20. Four short clamp bands 2 were fixed on metal angles 19 screwed to the base of the vibration installation. The clamp bands hold a support 11 with a special lower head 8. Ends of a glass pipe 9 with the length L = 918 mm, the inner diameter of d = 10.05 mm and the wall thickness of 2 mm were glued hermetically in an opening of the upper head 10 and the lower head 7. The pipe was held vertical with long clamp bands 8 fixed on the lower head 7 and the upper head 10.

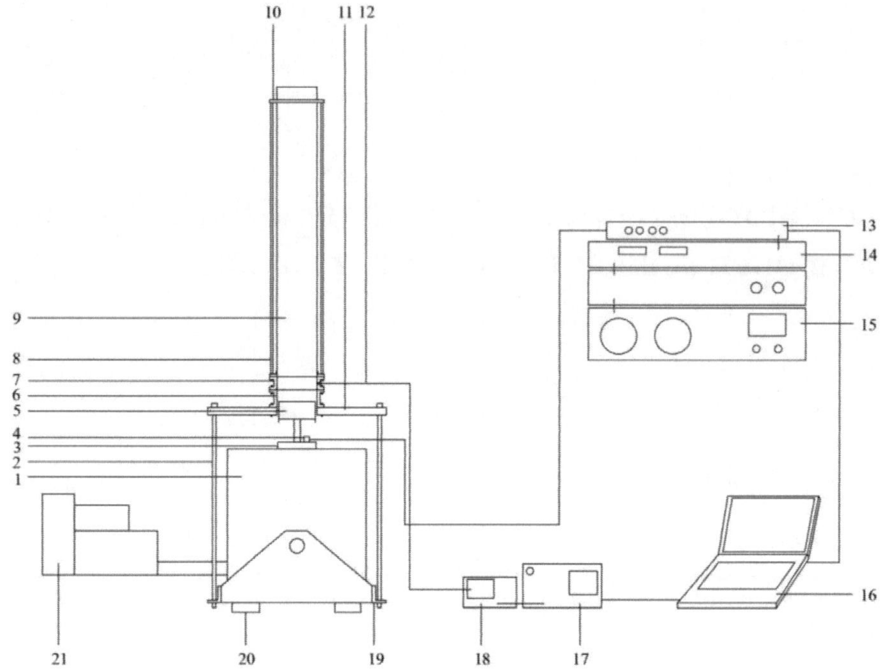

Figure 1. 1 – vibration generator, 2 – short clamp band, 3 – vibration table, 4 – piezoelectric IEPE accelerometer, 5 – flat piston, 6 – cylinder, 7 – lower head, 8 – long clamp band, 9 – glass pipe, 10 – upper head, 11 – support, 12 – pressure sensor, 13 – controller, 14 – positioning controller TMS-2, 15 – power amplifier, 16 – control computer, 17 – oscilloscope, 18 – three-channel bridge voltage amplifier, 19 – metal angles, 20 – pneumatic spring cushions, 21 – cooling fan.

The vibration table was controlled via a computer 16 by means of a piezoelectric IEPE accelerometer 4 of 4513 model by Bruel & Kjaer firm and a feedback controller 13 of VR9500 type by Vibration Research Corporation firm using a specialized VibrationVIEW software. The positioning controller TMS-2 14 was used to adjust the table of the vibration generator. A pressure sensor 12 of 8530C-15 model by Bruel & Kjaer firm was screwed into an opening of the head 7, and the signal from the sensor was fed through a three-channel bridge voltage amplifier 18 of ENDEVCO model 136 by Bruel & Kjaer firm to a digital oscilloscope 17 of DSO 3062A model by Agilent Technologies firm, and then via an RS-232 interface to the computer, where the data for the gas pressure p in bar can be observed and saved using a specialized DSO3000 software. A cooling fan 21 was used to prevent overheating of the vibration coil and the field excitation coil.

Sine oscillation of the piston near the first proper frequency ν_1 was set with the controller 13. Upon achieving the frequency and span of the piston oscillation $2l$ (where l is the span amplitude) with a certain value, a record of the pressure oscillogram was performed on the computer monitor. The voltage value ΔU in mV with the accuracy of 0.3 mV was converted into the value of the pressure oscillation span in bar using the conversion ratio ΔP (bar) = 0.00025 (bar/mV) × ΔU (mV).

2. Experimental results

The Figure 2 represents oscillograms with records of the gas pressure oscillations in time at different dimensionless excitation frequencies $\bar{\nu} = \nu/\nu_1$ and the piston span amplitude l = 0.75 mm. The transition through the resonance near the observed first proper frequency ν_1 = 90.3 Hz was observed. It is obvious that away from the resonance ($\bar{\nu}$ = 0.89), the gas oscillates according to the harmonic law. Along with further increase of the frequency ($\bar{\nu}$ = 0.98, 0.99), a distortion is formed and the amplitude increases in the discharge area. An increase of the steepness of the rising edge of the pressure wave in observed. At the resonance ($\bar{\nu} = 1$), the amplitude becomes maximum. After the resonance ($\bar{\nu} = 1.01$), the amplitude decreases, a further increase in the frequency ($\bar{\nu} = 1.06, 1.1$) leads to the fact that the gas oscillation once again assume a harmonic nature.

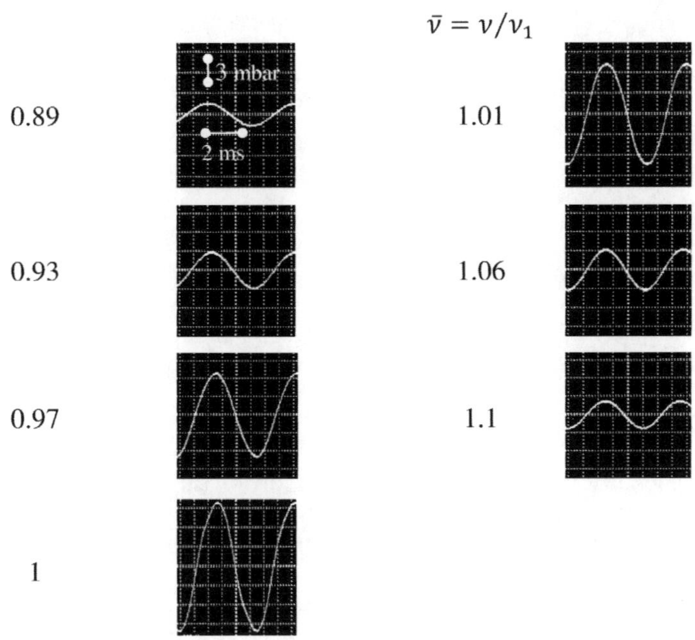

Figure 2. Oscillograms of the gas pressure oscillation at the amplitude of the piston shift l = 0.75 mm at different dimensionless excitation frequencies $\bar{\nu}$ near the first proper frequency $\nu_1 = 90.3$ Hz.

Quantitative results for the specified case are represented in Figure 3, where the amplitude-frequency curve is presented in the dimensionless form $\Delta\bar{p} = 10^2 \cdot \Delta p/p_0$ (p_0 – atmospheric pressure, $\Delta p = p_2 - p_1$ and p_1 are maximum and minimum values for the period of the piston oscillation). Notably, the value of the gas pressure amplitude at the resonance frequency is 4 times as high as away from the resonance. The theoretical resonance frequency is calculated according to the following formula

$$\nu_1^* = \frac{c_0}{2L(1+\beta')}, \qquad \text{where } \beta' = \frac{1}{2}\left(1 + \frac{\varkappa - 1}{\sqrt{\mathrm{Pr}}}\right)\frac{\delta}{R} \tag{1}$$

where c_0 is equilibrium sound velocity, and β' is absorption factor, $\varkappa$ is the adiabatic index, Pr is Prandtl number, δ is the acoustic interface thickness, R is the pipe radius, L is the pipe length.

The solid line marks the ratio of the excitation frequency to the first of the observed proper frequency, the dashed line marks the ratio of the excitation frequency to the proper frequency, calculated in consideration of the absorption according to the formula (1).

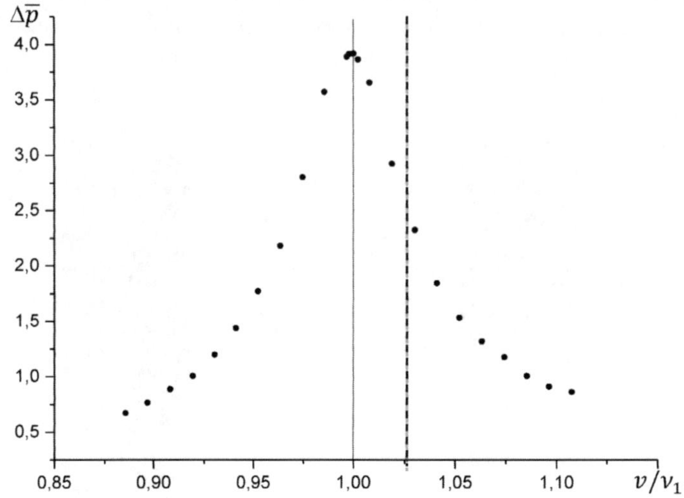

Figure 3. Dependency of the dimensionless span of the gas pressure oscillation on the dimensionless excitation frequency at the dimensionless amplitude of the piston shift l = 0.75 mm. Dots mark the experimental results.

4. Conclusion.

Peculiarities of process non-linear gas pressure oscillation in an open pipe in a wide range of frequencies. The deformation of the wave form observed: the steepness of the front rising edge becomes greater than that of the back one when approaching the resonant frequency. Pressure fluctuations have a continuous character. These new data on the characteristics of nonlinear oscillations of gas can be a good basis for increasing the efficiency of existing and to create new and promising applications. Since the nonlinear oscillations of gas may be a shock wave, the average flow vortex structures. Insufficient consideration of these effects in the development of technical devices can lead to a sharp reduction in the timing of their operation or an emergency.

References

[1] Ilgamov M A, Zaripov R G, Galiullin R G, Repin V B 1996 *Applied Mechanics Reviews* **49**(3) 137-154

[2] Saifutdinov A I, Fadeev S A, Saifutdinova A A and Kashapov N F 2015 *JETP Lett.* **102** 637–642

[3] Zaripov R G, Galiullin R G and Galiullina E R 2001 *Actual problems of continuum mechanics. On the 10th anniversary of IME KazSC RAS* 19-35 (Kazan: IME KazSC RAS)

[4] Zaripov R G, Tkachenko L A and Shaydullin L R 2014 *J. Phys.: Conf. Series* **567** 012042

[5] Gubaidullin D A, Zaripov R G, Tkachenko L A and Shaidullin L R 2015 *J. of Engin. Physics and Thermophysics* **88**(4) 843-847

[6] Saifutdinov A I, Saifutdinova A A, Kashapov N F and Fadeev S A 2016 *J. Phys.: Conf. Ser.* **669** 012045

[7] Zaripov R G, Kashapov N F, Tkachenko L A, Shaydullin L R 2016 *J. Phys.: Conf. Series* **669** 012053

Non-contact optical Liquid Level Sensors

Kiseleva L.L.[1], Tevelev L.V.[1], Shaimukhametov R.R.[1,2]

[1] LLC "INFORMATEKA"
[2] Institute of Engineering, Kazan Federal University

ramil.shai@mail.ru

Abstract
Information about characteristics of the optical liquid level sensor are present. Sensors are used to control of the light level limit fluid - water, kerosene, alcohol, solutions, etc. Intrinsically safe, reliable and easy to use. The operating principle of the level sensor is an optoelectronic infrared device.

Among the various liquid level sensors (capacitance, mechanical, microwave, optical, ultrasonic and others) optoelectronic devices have a special place [1, 2]. They are hardy to vibration, have no moving parts, work in all liquids, have small size and weight, durable, perfect in sanitary and ecological aspects, easy to operate.

Today, relatively simple optics in Russia is not as expensive as in Europe, and so optoelectronic devices is not more expensive (and in many cases, cheaper) devices other systems that allow them unconditionally to find its niche in all applications where their technical features and benefits evident.

Level switch designed to limit registration of liquids - water, ethyl alcohol, alcohol-water mixtures, light oil products (kerosene, gasoline, diesel fuel, etc.), liquefied natural gas and other fluids, non-corrosive materials for wetted parts (stainless steel, silicate glass, epoxy and polyester compounds)[3].

Design and coverage indicator provides resistance to oils and detergents.
The indicator provides protection level "Intrinsically safety"(with power supply units MIZ-460, UPS-8, UPS-300-2k and the like).

The operating conditions :

- Temperature of a controlled fluid, C .... -50 ... +60
- The ambient temperature, C .... -50 ... +60
- The difference between the ambient temperature
- and the controlled liquid, not more C 30
- The pressure in the reservoir at a controlled liquid, MPa not more 25
- Atmospheric pressure, MPa .... 0.084 - 0 .104

The operating principle of the level sensor [4] is a threshold photometer in visible or near IR range, detecting at the interface between two media with different refractive indices:
- The liquid phase - not less than 1.3;
- For the gas phase - not more than 1.1.

Table 1. Specifications.

	Unit of measure	Norms
The variance of alarm status in the whole range of external conditions, not more than	mm	1.5
The hysteresis not more than	mm	1.0

IOP Conf. Series: Materials Science and Engineering **134** (2016) 012023 doi:10.1088/1757-899X/134/1/012023

The switching time is not more than	ms	1
Supply voltage	V	6…24
Mean time between failures with the maintenance of at least	h	25000
Continuous mode		no time limit
Operating position		any
Connection thread		M20x1.5
Dimensions	mm	∅42x63
Weight, not more than	g	200

Characteristics of the sensor is Z-shaped. The transition from of logic "0" to "1" occurs when the liquid level is exceeds the level of alarm status.

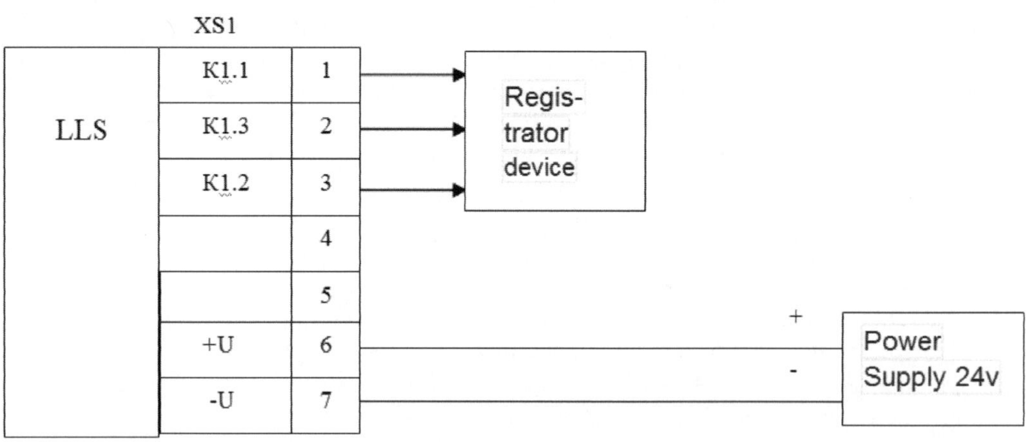

Fig. 1. Scheme signaling connection

In normal operation, indicator no requires maintenance during its lifetime.
The sensors operate in different industries [5] and has the following advantages:
Efficiency in tanks with pressure up to 50 mPa (various hydraulic drives systems), small size and weight, ease of application compared with other types of detectors (for metallurgy and other heavy industries).
The efficiency of the specific conditions (even in the flooded state (!) - Degree of protection IP-67 devices).The unique devices operating in the liquefied petroleum gas (LPG) in the vicinity of the heater power up to 5kW (for the energy sector).
The lack of direct contact of the device with a controlled fluid that ensures absolute cleanliness and safety control method, high accuracy and repeatability of registration of liquid level (up to 0.1mm). The unique devices working in the most demanding environments, such as a liquid oil, aggressive domestic and industrial waste water, foods with high visc metallurgy and other heavy industries osity and stickiness, including at high temperatures (in the food industry, pharmaceutical industry, biotechnology industry, medicine).

Small consumption of devices (5 ... 10 mA at a voltage of 6 ... 12 V), which greatly facilitates the construction where necessary explosion protection, performance in a wide range of climatic conditions (from -50 to +70 deg. C), long life (at least 50,000 hours).
Thus, the detector have a unique set of useful features that manifest in different fields of application.

Fig. 2. Sensor Versions.

References

[1] E D Tsamis and J N Avaritsiotis 2005 Sensors and Actuators A Physical **118(2)** 202-211
[2] M G Fazlyyyakhmatov N F Kashapov, Kh A Khayritonov D K Lazarev V K Lazarev 2015 IOP Conf. Series: Materials Science and Engineering 86 012020
[3] Huaping Gonga Haifeng Songa Sulei Zhangb Kai Nia Xinyong Dong 2014 Sensors and Actuators A: Physical Volume **205** 204 – 207
[4] Jon R. Askim Morteza Mahmoudi and Kenneth S. Suslick 2013 Chem. Soc. Rev. 42 8649-8682
[5] Baoquan Jin, Xin Liu, Qing Bai, Dong Wang, Yu Wang 2014 Measurement and Control - (MEAS CONTROL-UK), **47(7)** 219-224
[6] Kashapov L, Kashapov N and Kashapov R 2013 J. Phys.: Conf. Ser. **479** 012005
[7] Israfilov Z K and Kashapov N F 1991 Journal of Engineering Physics **60** 364–368

IOP Conf. Series: Materials Science and Engineering **134** (2016) 012024 doi:10.1088/1757-899X/134/1/012024

Two-layer flow of polymer melts in extruder die channel

R F Sharafutdinov [1], B A Snigerev[2], E R Galimov[1] and N Ya Galimova[1]

[1]Kazan National Research Technical University named of A. N. Tupolev-KAI, Russia
[2]Institute of Mechanics and Engineering, Kazan science center, Russian Academy of sciences, Russia

E-mail snigerev@imm.knc.ru

Abstract. The paper discusses numerical modeling of two-layer flow of viscous non-Newtonian fluids in extruder die channels. Fluid motion is described by mass and momentum conservation equations supplemented by the rheological equation of state of a viscous non-Newtonian fluid according to the Carreau model. Technique of numerical solution of the problem based on the finite element method is presented. Distribution pattern of fluid velocities, pressure, stresses, positions of the interface in the two-layer flow depending on the rheological properties of a fluid and flow regimes is investigated.

1. Introduction

Many products in the chemical, pharmaceutical, food and other industries are manufactured using multilayer materials. Lately, on the market of multilayer products the requirements to their consumer properties such as material quality, uniformity of layer thickness, density and others have increased. Thus, recently, there is much tension around the problem of improving the quality of technologies of multilayer coating on various surfaces. Investigation of two-layer flows of viscous non-Newtonian fluids in the channels will allow better defining the necessary parameters of technological processes of manufacturing the multilayer products [1]. The laminated or stratified flow of liquids has long attracted the attention of researchers. Very important are the studies related to the transportation of crude oil through pipelines. There are ways to reduce the pressure required to pump oil through the pipes, adding oil to the polymer solution, which attendees to the pipe walls, forming a two-layer flow. In a number of theoretical research to analyze the causes of the instability of shear flow of two-layer systems consisting of elastic viscous liquids, where the rheological equations of state used equations of non-Newtonian fluids are carried out [1,2]. The instability of stratified flow is expressed through fluctuations in the coordinates of the surface section relative to the flow axis. It is shown that the unstable shape of the interface surface is characteristic for flows which have substantially different viscosities and normal stress components at the line of interface. But most actively stimulated development of the stratified flow the problems encountered with co-extrusion method, when two or more streams of melt press through the forming head in common arising by production of layered polymeric films and biocomponent fibers [3].

IOP Conf. Series: Materials Science and Engineering **134** (2016) 012024 doi:10.1088/1757-899X/134/1/012024

In this paper the steady immiscible two-layer flows of non-Newtonian fluids formed as a result of the two flows coalescence in a plane channel with the formation of flow interface is investigated.

2. Governing equations

We consider nonisothermal flow without waves of two immiscible fluids in a converging channel; its scheme is shown in figure 1. The first fluid is supplied at the inlet Γ_1^1 (IH) with a flow rate Q_1, the second fluid is supplied at the inlet Γ_1^2 (AB) with a flow rate Q_2 after the flows convergence in EOL cross-section a multiphase flow with flows interface Γ_6 (OG) is formed. Area of fluid flow in the lower layer IHOGM is denoted as Ω_1 and in the upper layer BAEFGO – Ω_2 respectively. The flow of non-Newtonian fluids in each of the areas $\Omega_{1,2}$ is described by the equation for conservation of momentum and continuity equation in the form [2]

$$\rho^m\left(\frac{\partial u_i^m}{\partial t}+u_j^m\frac{\partial u_i^k}{\partial x_j}\right)=\frac{\partial \sigma_{ij}^m}{\partial x_j}, \tag{1}$$

$$\frac{\partial u_i^m}{\partial x_i}=0,\quad i,j=1,2,\quad m=1,2 \tag{2}$$

The system of equations (1) – (2) is closed by the rheological equation of state of a viscous non-Newtonian fluid

$$\sigma_{ij}^m=-p^m\delta_{ij}+\tau_{ij}^m, \tag{3}$$

$$\tau_{ij}^m=\eta^m(I_2)D_{ij},\quad D_{ij}=\frac{1}{2}\left(\frac{\partial u_i}{\partial x_j}+\frac{\partial u_j}{\partial x_i}\right),\quad \eta^m(I_2)=\mu_0^m(I_2)^{\frac{k-1}{2}}$$

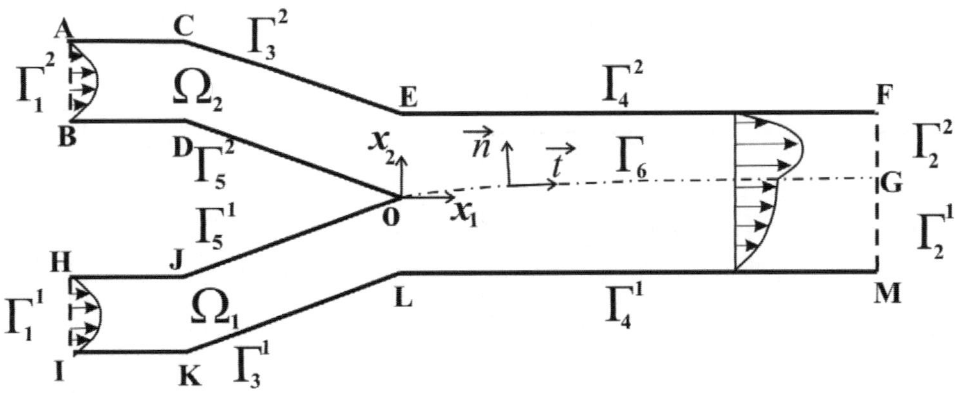

Figure 1. Scheme of flow of polymer melt in channel.

In (1) – (3) x_i – Cartesian coordinates, u_i^m – velocity components for a fluid with an index m, p^m – pressure, ρ^m – density, σ_{ij}^m – components of the compete stress tensor, δ_{ij} – components of stress deviator tensor, D_{ij} – components of strain velocity tensor, δ_{ij} – components of identity tensor, I_2 – second invariant of the strain velocity tensor, $\eta^m(I_2)$ – effective fluid viscosity, μ_0 – fluid consistency index (reduced viscosity), k – model parameter.

At the initial time the layers interface Γ_6 has a plane horizontal form. The axial velocity profile characteristic for the steady flow of non-Newtonian fluid in a plane channel is given at the inlet sections Γ_1^m. The no-slip conditions are satisfied at the solid walls $\Gamma_{3,4,5}^m$ and conditions of steady flow are set at the outlet boundaries Γ_2^m. Kinematic conditions must be met at the moving interface of the fluid layers Γ_6. In addition, dynamic conditions expressed in the equality of fluid shear stresses at each side of the interface must be met and normal stresses have a gap [4]. In two-dimensional case the interface surface is described by the equation $F(x_1,x_2,t)=0$, then the following conditions is fulfilled [3]

$$\frac{\partial F}{\partial t}+u_i^m\frac{\partial F}{\partial x_i}=0 \qquad (4)$$

Approximation of the equations (1) - (4) and calculations are carried out by finite element method of second order with linear and quadratic approximations of the unknown functions [5-7].

3. Results

We consider a two-layer flow of non-Newtonian fluids in a channel with a width in the area of multiphase flow of $h=2\cdot10^{-3}$ m, at flow rates of $Q^1=5\cdot10^{-9}$ м^3 /c, $Q^2=20\cdot10^{-9}$ м^3 /c. Values of the physical constants for the flow components are as follows: $\rho^1=930$ kg/m^3, μ_0^1 = 10010 Pa s, k^1 = 0.75, $\rho^2=980$ kg/m^3, μ_0^2 = 10050 Pa s, k^2 = 0.75.

In case of two-layer flow the interface position is determined by the flow rate values and their ratio as well as the ratio of melts viscosities of flows [1]. We denote by l^1, l^2 – the flow widths of the lower and upper layers respectively ($l^1+l^2=h$). We introduce the dimensionless quantity $h_f = l_2/(h/2)$, showing the degree of variation in the flow width at multiphase flow in the channel. The flow rate influence on the media interface position is shown in figure 2a which shows the distribution of h_f for various flow rate ratios Q^1/Q^2: *1* - Q^1/Q^2 = 5, 2-10, *3*-20. Values of h_f for the steady flow are h_f = 1.29, 1.73, 1.96.

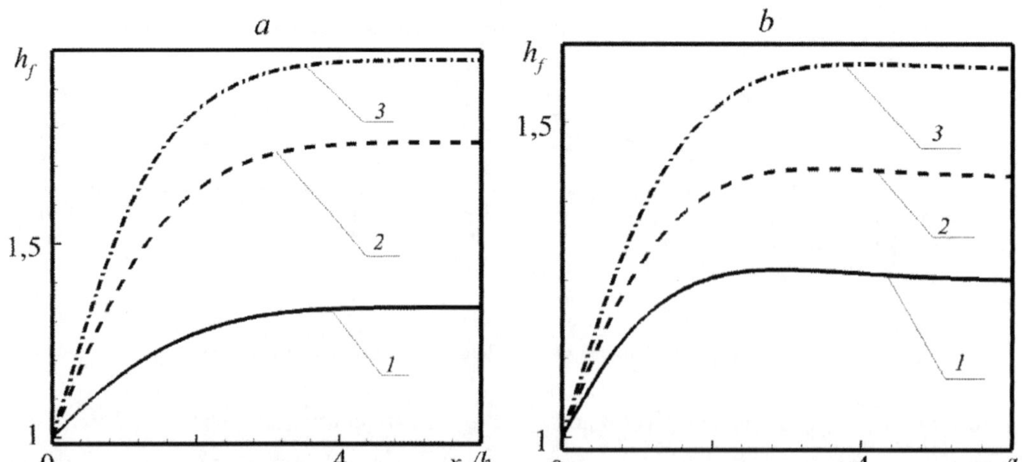

Figure 2. Profiles of liquid interface for different regimes.

The flows were calculated under condition of constant flow rates at the lower and upper layers $Q^1=Q^2$ to determine the influence of different viscosity on the interface position. The viscosity influence was determined under the conditions when the lower layer viscosity remained constant μ_0^1 = 10010 Pa s, k^1 = 0.75, the upper layer coefficient μ_0^1 increased and the ratio μ_0^2/μ_0^1 was equal to μ_0^2/μ_0^1 = 5, 10, 20. Since the fluid viscosity of lower layer in the area of the multiphase flow is higher than in the upper layer this leads to a shift of the interface position from the channel axis to the upper wall. Further increase of μ_0^2/μ_0^1 leads to the growth of deviation as illustrated in figure *2b*. This figure shows the h_f distribution for three viscosity modes: 1- μ_0^2/μ_0^1 = 5, 2- 10, *3*-20.

In the present paper the flow structure and the stress-strain state of the media at the two-layer flow of viscous non-Newtonian fluids in the channel are investigated theoretically. Significant influence of difference in values of fluid flow rate ratios, viscosity consistency indexes and pseudoplasticity degree of non-Newtonian media on the interface position is shown. The influence of the measure of discrepancy of the flow rates, viscosities and non-Newtonian properties of pseudoplasticity is investigated.

References

[1] Yankov V I and Glot I O and Trufanova N M and Mitrofanova OV 2010 *D Polymers flow in the die holes. Theory, calculations, practice.* Editorial URS Izhevsk
[2] Chang D H 1979 *Rheology in polymer Processing* M.: Chemistry
[3] Volodin V P 2005 *Extrusion of profiled products from thermoplastics* Saint-Petersburg: Publishing house Professiya
[4] Connor J and Brebbia C 1979 *Finite element technique for fluid flow* L: Sudostroenie
[5] Sharafytdinov R F and Galimov E R and Snigerev B A 2015 *IOP Conf. Series. Materials Science and Engineering* **86** 012031
[6] Gubaidullin D A , Morenko I V and Fedyaev V L 2015 *Russian Aeronautics (Iz VUZ)* **58** 351–354.
[7] Saifutdinov A I, Saifutdinova A A, Kashapov N F and Fadeev S A 2016 *J. Phys.: Conf. Ser.* **669** 012045
[8] Saifutdinov A I, Fadeev S A, Saifutdinova A A and Kashapov N F 2015 *JETP Lett.* **102** 637–642

IOP Conf. Series: Materials Science and Engineering **134** (2016) 012025 doi:10.1088/1757-899X/134/1/012025

On the mechanism of formation of wear-resistant coatings on the friction surfaces of technical products in the presence of these drugs Tribo

S N Sharifullin[1], **A V Dunayev**[2]

[1] Kazan Federal University, 420008, Kremlyovskaya st., Kazan, Russia

[2] All-Russian Research Institute of Technology repair and maintenance tractor fleet (GOSNITI),109428, 1 st Institutskiy passage, house1, Moscow, Russia

E-mail: Saidchist@mail.ru

Abstract. Extending the service life of technical devices by exposing the surface of the friction tribo different drugs is an established fact. There are various hypotheses to explain the mechanism of formation of coatings with high wear resistance and low coefficient of friction on the mating surfaces of friction units of machines and mechanisms during their operation with the presence of drugs Tribo. All these conflicting hypotheses. The proposed technology without the wear operation of equipment using Tribo drugs differ in content and methodology of processing machinery. In this paper, an analysis of existing hypotheses formation mechanism of these coatings. The authors have their own experience in the development and application of technologies without the wear operation of equipment using Tribo drugs. This allows them to develop their own hypothesis of the mechanism of formation of coatings with high wear resistance and low coefficient of friction on the mating surfaces of friction units of machines and mechanisms during their operation with the presence of drugs Tribo.

1. Introduction

In the operation of any technical product having a friction units, as wear parts loses its efficiency and as a result is unable to perform specified functions with the parameters set by the requirements of the technical documentation. Today, all over the world have been exhausted technological possibilities of manufacturing different machines and mechanisms without noticeable wear the mating surfaces of friction units. Lubricants used to reduce friction, and have exhausted the limits of its perfection. In addition, they do not eliminate wear mates. Steel is used for modifying the Tribo environment, so-called additives. Purpose additives - Improving the quality of lubricants and, primarily, to improve their antiwear and antifriction properties. They are based on high molecular weight polymers, Detergent and dispersant components, oxidation inhibitors, rust and corrosion, anti-silicones or poliloksany. However, additives allow only modify Tribo environment with improved lubrication properties. When changing the lubricant additives require re-application.

Almost impossible to eliminate wear. Therefore the search for unusual ways to prolong the service life of technical devices by acting on the friction surface is an urgent task. Currently, one such method is known. It is based on the exploitation of the machinery to the presence of friction tribo different drugs. This method allows, under certain conditions actually produce the so-called operation without the wear art. The essence of the technology without folding restore worn surfaces of friction units of machines and mechanisms is that RVS additive introduced into regular oil or grease. The preparation is then transported oil on the surface of the friction units, and under the influence of pressure, temperature and and tribes plasma creates conditions for the return of wear on the surface of the worn-out place with a partial or full recovery of the geometric dimensions of parts [1-7]. Thus the formation of layers with high wear resistance and low friction.

2. Results and discussion

In the world there are different Tribo drugs to prolong the service life of machinery. These preparations can be organic and inorganic nature. Preparations inorganic nature called tribo additives or modifiers, and organic nature - geomodifiers or mineral friction modifiers. Unlike additives tribo preparations of both organic and inorganic nature on the friction surfaces of parts creates notable protective film [1]. Preparations of inorganic nature on part surfaces form a short-lived anti-friction coatings, and organic - long-lasting anti-friction coating. When changing the lubricant reapplication Tribo preparation is required. Studies have shown that the anti-friction coating of organic products Tribo close to diamond-like coatings. Their hardness is 70 HRC, wear resistance several times higher than the wear resistance of the original surface of the workpiece, and the coefficient of friction is reduced by more than 2 times, and sometimes up to 90%.

Since the 60s of last century began to emerge new scientific and technical sphere - management of friction, wear and restoration of worn surfaces during the operation of machines and mechanisms using artificially caused by physico-chemical phenomena in friction, manageable ntroduction to the tribological interface products with feedback on operating parameters of interfaces. Education layers with high wear resistance and low coefficient of friction on the mating surfaces of friction units of machines and mechanisms during their operation is an established fact. Therefore, at the end of the last century, manufacturers Tribo drugs, both in our country and abroad, technology without collapsible rebuilding worn friction units of machines and mechanisms using geomodifiers steel used in commercial purposes. This technology is based on the so-called opening "effect of low friction surfaces of friction units in the presence of phyllosilicates." The availability of this technology, simplicity, low cost and economic benefits allowed its widespread development of different kinds of practices, experimenters. But it almost to the present day has not been studied mechanism of wear-resistant coatings on the surfaces of friction units of machines and plants in the presence of Tribo preparations not performed indepth research on the study of processes taking place at the same time. Therefore, not otrabotanom this technology, its use of "blind", and sometimes led to negative results. Hence some caution in the technology and the lack of its large-scale application [8].

In recent decades, there have been some studies that reveal the essence of the processes taking place in the friction units of machines and mechanisms in the presence of these drugs Tribo. It turned out that these processes play a major role in the resulting contact points mating surfaces friction unit electrical discharges, was later called "Triboplasma" [9]. Existing studies have shown that an electrical discharge that occurs in friction machines and equipment relates to the section of low-temperature plasma. Then triboplasma equations can be described as applied to lowtemperature plasma. But here it is necessary to consider that the existence of a single spot on the mating surfaces of the friction unit is about $10^{-8} - 10^{-7}$ c. This further complicates the development of the theory of the mechanism of action of repair and restorative composition (RVS) on the surface of the friction units, which is the main hindrance RVS optimization technology, and its large-scale deployment in the industry and repair production.

There are various hypotheses to explain the mechanism of formation of coatings with high wear resistance and low coefficient of friction on the mating surfaces of friction units of machines and mechanisms during their operation with the presence of drugs Tribo. All these conflicting hypotheses. The proposed technology without the wear operation of equipment using tribo drugs differ in content and methodology of processing machinery.

To date, there are about ten hypotheses on the mechanism, sometimes conflicting. Their different understanding and ambiguity related to the lack of a clear understanding of the processes occurring in the friction units of machines and mechanisms with the presence of drugs Tribo. Therefore, it is planned to develop its own hypothesis, the most suitable for the actual mechanism of formation of wear-resistant coatings. For clarity, we present some common hypothesis on the subject [8]:

- One of the first in time and the most common - a "technological" hypothesis of the NGO "Ruspromremont", ie cold deformation, which is understood as a mechanical pressing onto the friction surfaces of products crushing and destruction of minerals. The advantage of this hypothesis - keeping the pressure as one of the main factors of the recovery process of the friction surfaces HMT-additives. However, this hypothesis does not explain the complex structure of tumors, significantly different from that of the locus;

- Other pioneers RVS technology with physical positions assumed that the restoration of the worn surface friction and increasing the protective layer, offset by increased gap interfaces friction is the result micrometallurgical processes in local points of frictional contact. The advantage of this hypothesis - account other important factors tribohimii - the temperature in point. However, the evidence of the elemental composition HMT-coating shows the presence of a predominant (70 ... 85%) of the amount of carbon, a small (3 ... 5%) and similar to the amount of iron - silicon, in the absence of appreciable amounts of magnesium, which does not confirm the accuracy of the cermet ;

- In NPIF "anyon Baltica" an important mechanism for the diffusion of the work locus believed some of its components (for example - magnesium) into the metal surfaces of the parts. Hence the name given tribopreparatam "anyon-Baltic": NIOD (focused ion diffusion). This hypothesis requires confirmation CONTENTSon high magnesium surfaces of parts after-treatment of PBC. However, our studies have shown that magnesium is absent in the coatings. In our opinion, the only magnesium plays a catalytic role in the ongoing processes in friction;

- SPC "Converse Resource" explains its action tribo drugs based on copper salts of fatty acids and serpentine from the standpoint of "selective migration". However, rarely implemented selective transfer occurs only under strictly defined combination of materials of friction pairs and only the friction of steel on copper alloys;

- Developers serpentine preparation "SUPROTEC" believe that the active components are formed on the surface of friction cellular structures, retaining grease. These structures form the surface, which the authors called for the preparation triboparah "third layer of the body." They believe that at the beginning of the drug under the influence of high pressure and temperature in the ultrafine powder mates formed iron carbides, which removes the friction surface contamination, and then the surface is covered with juvenile cellular structure;

- Ph.D. Rybnikov Yu, an expert on the glass for the mechanism of recovery of worn surfaces with HMT offered close to the hypothesis sintering secondary structures, the formation of the hypothesis "the iron-magnesium glass." A small presence of iron in the structure of PBC coating explains its high adhesion to the surfaces of parts (only iron-carbon alloy) reaction with iron atoms of the friction surface. With the same hypothesis "vitrification" RVS surfaces consistent high grade purity of the newly-formed film, its high hardness, high electrical resistance, heat resistance, and transparency. But this hypothesis does not explain maslofilnost film and it is dominated by the presence of carbon, not hypothetical glass components;

- The justification of an entire era of crystal-chemical mechanisms work and how to use minerals as friction modifiers are works of one of the pioneers of the RVS technology, Dr. geol.-miner. Sciences, Academy of Natural Sciences VV Zueva;

- Noteworthy recent physicochemical presentation Ph.D. DM Teluha. He states that there are five basic types of serpentine: antigorite, chrysotile, klinohrizotil, ortohrizotil and lizardite. But, according to him, most of the RVS-composition prepared from the lizardite doped with magnetic and chrome iron ore particles of olivine, which determines the instability of the compositions and the variation of their tribological properties - with microimpurities talc, silica and other unwanted clay minerals, oxides of aluminum and silicon, non-adhesion to steel.

IOP Conf. Series: Materials Science and Engineering **134** (2016) 012025 doi:10.1088/1757-899X/134/1/012025

In this issue there and our preliminary hypothesis, from our point of view, the most real. It is as follows. From the set of considered viewing versions can be concluded that the major factors of the process are: the pressure and temperature in a local point tribocontact, the presence of a catalyst element (magnesium) in tribo preparation, a large amount of heat (energy) emitted in the destruction of mineral friction modifiers node contact friction surfaces , triboplasma, grinding particles with their abrasive effect on the unevenness and oxide films, the presence of oxidation products tribosrede organic lubricant, polycondensation, tribopolimerizatsiya and adhesion to the friction surfaces of some modified components of the friction of the friction environment, diffusion in the surface friction others chemisorption third.

In general, when the serpentine tribo drugs, early appears, it is cleaned from the friction surface layers, oxide films deformed and unstable crystal structures and cavitation abrasive mineral particles with the opening juvenile, catalytic surfaces. Juvenile surface - a newly formed surface of the part with the free molecular bonds and free of oxides and other contaminants. Only after grinding particles and the opening juvenile friction surfaces occurring serpentine particle conversion processes and modification of the friction surfaces.

3. Conclusions

Presumably, formation of a coating with high wear resistance and low friction takes place in stages: first there is treatment of industrial pollution, then the electrochemical and electrical processes, then the formation of an amorphous coating fragile, after maturation and hardening by peening.

References

[1] Dunaev A.V. Modernization of equipment worn with tribopreparatov. Monograph /A.V. Dunayev, S.N. Sharifullin. - Kazan: Kazan. University Press, 2013. – 272 p.

[2] Dounaev A, Sharifullin S 2014 *World Applied Sciences Journal* **31**(2) 272–276.

[3] Sharifullin S and Rogozhkin N 2014 *IOP Conference Series: Materials Science and Engineering* **69** 012039

[4] S.N. Sharifullin. Modernization of worn nodes and mechanisms of friction with nanotechnology / S.N. Sharifullin // Abstracts of the II International scientific and practical conference "Perspective materials and technologies for production" in the framework of the International Forum "Crimea Hi-Tech - 2014", Sevastopol, 22 - 28 September 2014. - M .: NRTU "MISA" , 2014. - Pp. 277 - 282.

[5] Saifutdinov A I, Fadeev S A, Saifutdinova A A and Kashapov N F 2015 *JETP Lett.* **102** 637–642

[6] Saifutdinov A I, Saifutdinova A A, Kashapov N F and Fadeev S A 2016 *J. Phys.: Conf. Ser.* **669** 012045

[7] Zaripov R G, Kashapov N F, Tkachenko L A, Shaydullin L R 2016 *J. Phys.: Conf. Series* **669** 012053

[8] R.Y. Solovyov Without the wear operation of internal combustion engines. Monograph /R.Y. Solovyov, S. Sharifullin, S.A. Solovyov, A.K. Olhovatsky V.B. Lomukhin, A.V. Dunayev, D.A. Gitelman, N.C. Hismetov. - M .: FGBNU GOSNITI, 2015. - 196 p.

[9] Kashapov N F and Sharifullin S N 2015 *IOP Conference Series: Materials Science and Engineering* **86** 012022

Dynamic model of production enterprises based on accounting registers and its identification

R T Sirazetdinov[1], A V Samodurov, I A Yenikeev[1], D S Markov[1]

[1.] Kazan National Research Technical University. AN Tupolev KAI

Abstract. The report focuses on the mathematical modeling of economic entities based on accounting registers. Developed the dynamic model of financial and economic activity of the enterprise as a system of differential equations. Created algorithms for identification of parameters of the dynamic model. Constructed and identified the model of Russian machine-building enterprises.

1. Dynamic economic and mathematical model of the enterprise

In this paper are reviewed, the problem of economic and mathematical modeling of industrial enterprises. The dynamic model of financial and economic activity of the enterprise was developed as a system of differential equations, which builds on the accounts. Algorithms for identification of parameters of the dynamic model have been developed. Work is carried out in line with the scientific directions initiated by academician of the Academy of Sciences of the Republic of Tatarstan T K Sirazetdinov at the Kazan Aviation Institute since 1970s.

In the framework these studies developed a general block diagram of the functioning enterprises based on accounting registers (Fig.1.) Principles of dynamic economic and mathematical modeling at the system of differential equations, algebraic equations and inequalities based on accounting records. [4] In this article the method of identification of dynamic models based on financial statements that are implemented in the construction of the business model of manufacturing enterprise.

On the structural model the accounts are grouped by their purpose and presented as an blocks of the model. Communications between the blocks defined by correspondence of accounts and presents with a physical or financial flows.

Each block is described by differential and algebraic equations and inequalities. The whole model is a system of differential and algebraic equations with variable structure and corresponds to a chart of accounts of the enterprise.

In the base of a mathematical model of the enterprise laid the mathematical model of accounting bills, which in algebraic form is described as follows:

$$S = S_0 + O_{in} - O_{out}\ , \qquad (1)$$

где, S – residue of resources at a certain time t;

S_0 – residue the resources at the initial time t_0;

O_{in} – the flow of resources in a time interval (t_0, t];

O_{out} – produce of resource in the time interval (t_0, t].

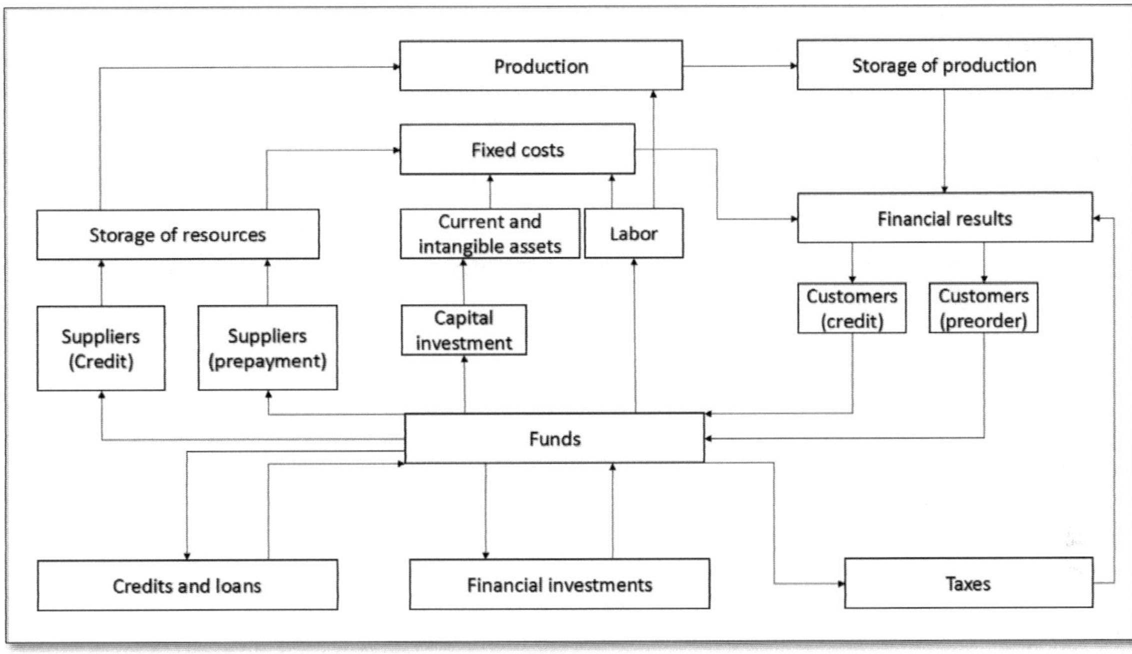

Figure 1. Block diagram of the enterprise based on a book of accounts.

The above presented formula (1) in differential form is written as follows:

$$\frac{dS}{dt} = o_{in} - o_{out}, \quad S(t_0) = S_0, \text{ with } S > 0;$$
$$o_{in} \geq o_{out}, \text{ with } S = 0; \tag{2}$$

Where, o_{in} – the flow of resources at time t (input);

o_{out} – the flow release resources at time t (output).

In some cases we have communication between input and output flows as a delay. For example, the amount paid to suppliers for material resources at some time t, the unit enters the "suppliers", and after a certain period of time τ, coming out of the unit at the same rate, only in the form of delivery. Blocks of this kind are described by the following equation:

$$o_{in}(t) = o_{out}(t - \tau). \tag{3}$$

where, τ –the average time of resource spent in the block, for example, the average time delay between the delivery of material assets and its payment, etc.

The above formula (3) represents a net delay, and to identify its parameters need to know the background of the process, which is not always possible. So often to describe the use of such a formula depending on the inertial delay:

$$\tau \frac{do_{out}}{dt} + O_{out} = O_{in};\ O_{out}(t_0) = O_{out_0}. \tag{4}$$

Thus, all blocks of the model may be of two types - type "warehouse" - expression (2), and the type of "delay" - the expression (4) is a differential equation. Links between blocks are described by algebraic equations.

For the identification of the model parameters used enterprise accounting data, because they are almost directly correspond to the blocks of the structural model and show sufficient detail the state of

IOP Conf. Series: Materials Science and Engineering **134** (2016) 012026 doi:10.1088/1757-899X/134/1/012026

the entity. However, some parameters of the model can not be directly derived from the financial statements and require special identification algorithms.

2. Identification of system parameters

The initial state of the system is determined by the balances at a given time t. As an example, such parameters include the balance of the financial result, the residue of finished products for sale as WIP residue etc.

Time parameters of the system, ie, delays associated with the time difference between input and output streams are identified by the formula of Professor Sherr:

$$\tau = \frac{\bar{S}\,P}{O}, \quad (5)$$

Where, $\bar{S}$ – the average balances on the account;

P – the duration of the analyzed period;

O – turnover for the period under review.

Average balances may be in the arithmetic mean formula

$$\bar{S} = \frac{\frac{S_{\text{н}}}{2} + \frac{S_{\text{к}}}{2}}{2} \quad (6)$$

Where, $S_{\text{н}}$ – the balance of the account (s) at the beginning of the reporting period;

$S_{\text{к}}$ – the balance of the account (s) at the end of the reporting period.

Temporal parameters of the system include, for instance, the time of production, time writing off fixed costs, the delay between payment to the supplier and obtaining resources, the delay between the receipt of resources and their payment, the delay between receiving the loan and the payment of the loan, etc.

Besides there are some system tuning parameters, which are values of these or other factors of production, which directly or indirectly affect the character of the movement of commodity and money flows. These parameters are on the one hand controls flow from another company reflect external factors affecting its operation.

The external factors include commodity prices, consumer purchasing power, taxation, etc. These values are generally set by experts. The internal factors can be attributed to the example of the economic policies in the enterprise. Shares of provided resources on credit and prepaid share of supplied goods on credit and prepaid production costs, etc.

Control these parameters and system solutions of equations in time, we get the options forecasting operation of the business in different types of external and internal factors.

3. Conclusion

Identification algorithm proposed by the authors, are implemented in Microsoft Excel environment with the use of the program on the algorithmic language VBA. A comparison of the predicted values of the parameters of the production facility with actual facts occurred. The results indicate quite close agreement of a number of identified parameters and their actual values.

4. References

1. Sirazetdinov T K *Dynamic modeling of economic objects* // Feng, 1996.

2. Sirazetdinov T K, Rodionov V V Sirazetdinov R T *Dynamic modeling of the region's economy*. Feng the Academy of Sciences, 2005.

3. Sirazetdinov R T, Brazhkina A A *Universal structural model of a typical economic cluster // Managing large systems*. Issue 29. M .: ISP RAS, 2010. S.152-166.

4. Sirazetdinov R T, Enikeev I A *The block diagram of a dynamic economic and mathematical model based on a book of accounts.* // "Society, the state, the personality: problems of interaction in a market economy." // VII Interuniversity Scientific Conference, Kazan, 2006. S.224-226.

5. Sirazetdinov R T, Enikeev I A *The main types of blocks used in the modeling process of the company.* // Innovations in science, technology, education and social services. // Proceedings of the All-Russian Youth Scientific and Practical Conference, 2003. P.57.

6. Sirazetdinov R T, Enikeev I A *The use of economic and mathematical modeling to the analysis of the production facility.* // First Spiridonovskie read "Actual problems of modern Russia's economy." // International Scientific Conference 2003 P.189-190.

7. Sirazetdinov R T, Enikeev I A. // 4th Spiridonovskie read *"Actual problems of modern Russia's economy."* // International Scientific Conference 2006 P.81-83.

IOP Conf. Series: Materials Science and Engineering **134** (2016) 012027 doi:10.1088/1757-899X/134/1/012027

Regasification of liquefied natural gas and hydrogen

V G Tonkonog[1], A L Tukmakov[1], K M Muchitova[1], U A Agalakov[1], F Sh Serazetdinov[2] and B C Gromov[2]

[1]Kazan National Research Technical University of A. N. Tupolev-KAI, Russia
[2]OOO NPP «Aviagas-Coyus+», Russia

E-mail tukmakov@imm.knc.ru

Abstract. Liquefied natural gas and hydrogen gasification process is suggested, in which vapor phase is generated by the decrease of internal energy of the liquid. Methane and hydrogen gasification processes have been numerically modeled. Flow rates of the methane and hydrogen through choke channel were defined. A satisfactory match between the modeled and experimental data for liquid nitrogen has been acquired. Technical suitability of the suggested process is proved. Based on the initial parameters of the cryogenic fluid, the amount of vapor phase is 5-20 % of the flow rate.

Liquefied natural gas (LNG) and liquid hydrogen (LH) are cryogenic fuels that are widely used in aerospace and ground power engineering. Lately LNG usage is growing in many countries [1]. LNG-based technological processes include transportation, as well as storage and vaporization. In example, LNG or LH is usually stored in tanks as liquid, and then vaporized prior to utilizing it in a power plant. Same occurs in local gas supply: LNG/LH is supplied to consumer as liquid, and vaporized prior usage. As following, development of vaporization technologies is most perspective.

Cryogenic fluid regasification might be performed in multiple ways: energy in the form of heat or mechanical work could be brought to it from outside environment, or LNG/LH can be vaporized using its own internal energy, or these ways could be combined.

Heating LNG by heat exchangers is highly feasible if vapor phase consumption rate is constant. However, when consumption rate is unstable, vaporization technology based on heat exchangers is somewhat limited due to heat exchangers are highly inertial. In this paper technological process of vaporization based on internal energy loss is proposed. Process is protected by RF patent № 2347934.

Proposed technological process consists of: cryogenic fuel storage tank, vapor phase generating choke, gas-liquid separator. Process stages are following: saturated liquefied cryogenic fluid is stored in heat-insulated tank at temperature To and pressure p which is saturation pressure for To ($p = p_s(To)$). Fluid is supplied to choke, either moved by pump or by vapor cap expansion in the tank. At choke fluid flows with negative pressure gradient, so its internal energy decreases and some amount of vapor phase is generated, so single-phase flow is transformed in two-phase gas-liquid flow, in which liquid phase temperature is lower than initial To because of internal energy loss. Then two-phase flow is separated in gas-liquid separator, liquid is returned to the storage tank and vapor is supplied to consumer. Since generated vapor and residual liquid temperature is lower than To, returned liquid phase cools fluid stored in tank, therefore improving gasification efficiency and decreasing the necessity of vapor drainage from tank. Proposed process has good dynamics and is easily controlled by changing choke cross-section.

In order to provide an estimation for real characteristics of proposed process, vapor generation by Laval nozzle was numerically modeled. Laval nozzle was selected as a choke because it has most feasible vapor generating characteristics. Both single-dimensional and two-dimensional cases of thermally non-equilibrated two-phase flow have been numerically modeled.

Single-dimensional model is based on following assumptions: flow is adiabatic, vaporization occurs in metastable part of phase envelope, liquid phase is overheated and amount of overheat is set according to [2], vapor phase is saturated, vapor and liquid phases have equal velocities. System of equations was solved using Runge-Kutta algorithm. Solution algorithm includes critical condition, at which flow rate through choke aspires to critical flow rate $J \rightarrow J_{cr}$ and pressure gradient is negative along all the choke $dp/dz<0$.

Calculated vapor mass fraction generated during H2 and CH4 adiabatic expansion in the Laval nozzle are presented on figure 1. Vapor fraction calculated both for equilibrated isentropic flow and non-equilibrated flow allows to estimate range of vapor fraction dependant on flow non-equilibrium rate. Results of calculation of vapor generator capacity for critical flowrates of N2, CH4 and H2 from state of saturation dependant on reduced pressure π are presented on figure 2, as well as R.J. Simoneau [3] and our own experimental data.

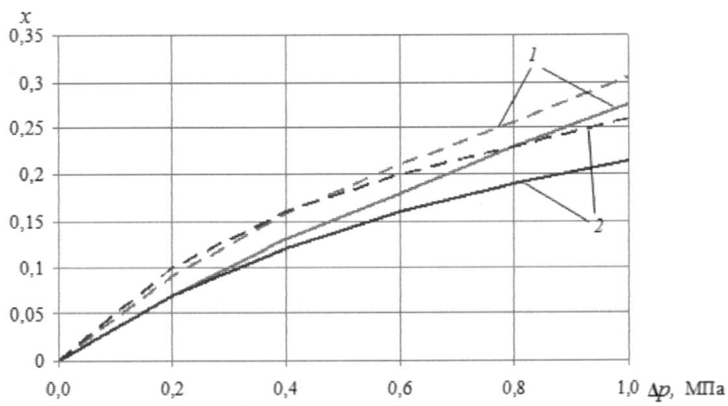

Figure 1. Vapor mass fraction, generated during adiabatic expansion in the choke channel. 1- normal H2, 2 – CH4.(----) – equilibrated process; (—) – non-equilibrated process

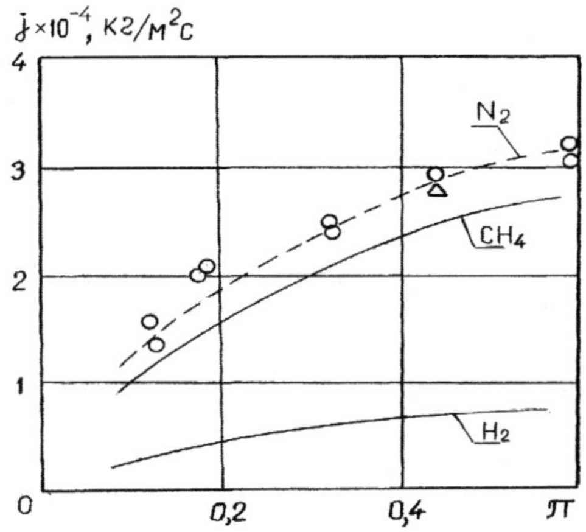

Figure 2. Specific flow rate through choke channel (Laval nozzle). ○-own experiment; Δ-R.J. Simoneau data [3].

IOP Conf. Series: Materials Science and Engineering **134** (2016) 012027 doi:10.1088/1757-899X/134/1/012027

Figure 2 shows satisfactory match between one-dimensional non-equilibrated model and experimental data, both our own and other researchers. Since fluid flow accompanied by its phase transitions has highly heterogeneous flow regime, for separator modules (which are important part of gasification process) development it is crucial to know local parameters of flow. Mainly it concerns local vapor fraction and flow regime. In order to solve this problem, the behavior of the two-phase flow in the separator for the two-dimensional case has been numerically modeled [4,5]. Te inversed problem has been solved – at given flow rate (calculated in one-dimensional case) separation characteristics have been simulated. Flow model and solution method is described in [6]. Figure 3 shows calculated vaporization front as well as vapor fraction distribution along the Laval nozzle for CH4 adiabatic flow.

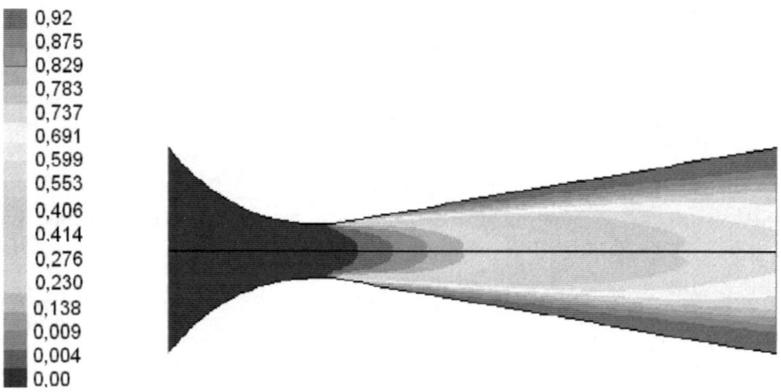

Figure 3. Vapor volume fraction for CH4 two phase adiabatic flow at initial parameters: p_o=0,5 MPa, T_o=130K

Cryogenic fluid regasification technological process is proposed. The H2 and CH4 gasification characteristics have been numerically simulated. Results might be used in cryogenic fluids gasification tools development.

Research was possible due to Ministry of Education of Russian Federation support as Federal Target Program (FTP). Grant contract № 14.577.21.0151. Topic: "Gasification process development and technology adaptation for power engineering", Project unique ID is RFMEF157714X0151.

References

[1] Vovk V S, Novikoff A I and Glagoleff A I etc 2009 *Liquid natural gas world industry and market: predictive modeling* M.: Gazprom expo

[2] Tonkonog V G and Gortyshov Yu F 2007 *Heat Transfer Research* **38** 661-668

[3] Simoneau J R 1975 Pressure Distribution in a Converging Diverging Nozzle During Two - Phase Choked Flow of Subcooled Nitrogen *In: ASME Non -Equilibrium Two -Phase Flow Symp* 37-45

[4] Saifutdinov A I, Saifutdinova A A, Kashapov N F and Fadeev S A 2016 *J. Phys.: Conf. Ser.* **669** 012045

[5] Saifutdinov A I, Fadeev S A, Saifutdinova A A and Kashapov N F 2015 *JETP Lett.* **102** 637–642

[6] Tonkonog V G and Bakoush A M 2006 *Izv. vuzov. Aviation tech.* **4** 47-49

IOP Conf. Series: Materials Science and Engineering **134** (2016) 012028 doi:10.1088/1757-899X/134/1/012028

Investigation the flushing flow of liquid methane in Laval nozzle

B A Snigerev[2], A L Tukmakov[1] and V G Tonkonog[1]

[1]Kazan National Research Technical University of A. N. Tupolev-KAI, Russia
[2]Institute of Mechanics and Engineering, Kazan science center, Russian Academy of sciences, Russia

E-mail snigerev@imm.knc.ru

Abstract. Turbulent flushing flow of methane in Laval nozzles are investigated. To describe the motion of vapor-liquid mixture are used Favre averaged over the set of equations including the equations conservation of mass, momentum, and energy for a homogeneous mixture consisting from liquid and vapour phases. Numerical flow simulation based on cavitation approach using an additional transport equation for the volume fraction of the liquid phase. The study of the expiry of boiling methane at different degrees of underheating and the back pressures in the Laval nozzle are performed.

1. Introduction

With the introduction of cryogenic temperatures in various industries are becoming increasingly important problems of efficient use of cryogenic products on an industrial scale [1]. At the present time in various fields of national economy: in mechanical engineering, metallurgy, rocket and space technology and others use liquid cryogenic products. Such products of gases under normal conditions, having found the greatest application, are methane, oxygen, argon, nitrogen, hydrogen and helium. The extensive use of these substances in a liquid state, due, on the one hand, the general progress of development of cryogenic technology: improvement of processes and equipment for liquefaction and separation of gas mixtures listed above products, as well as products for storage and transportation. In this regard, the need to create a liquid cryogenic systems, taking into account the specific properties of the products, in recent decades there have been numerous studies of individual processes that accompany the storage and transportation of fluids. The liquid methane is used as a liquid fuel. Its application can reduce the content of harmful additive in the exhaust gases at the same time improve the efficiency of engines. Methane - the main component of natural gas, which is widely used as a fuel and raw material for the chemical industry.

2. Governing equations

IOP Conf. Series: Materials Science and Engineering **134** (2016) 012028 doi:10.1088/1757-899X/134/1/012028

The flow of methane in the Laval nozzle, shown in figure 1 are considered. The front width of the nozzle is $l_1 = 96$ mm, the output width of a cut of the channel is $l_2 = 23$mm, in the throat portion is $l_4 = 4$mm. The length of the diffuser part equal to $l_3 = 150$ mm, and of the convergent channel is $l_5 = 50$mm.

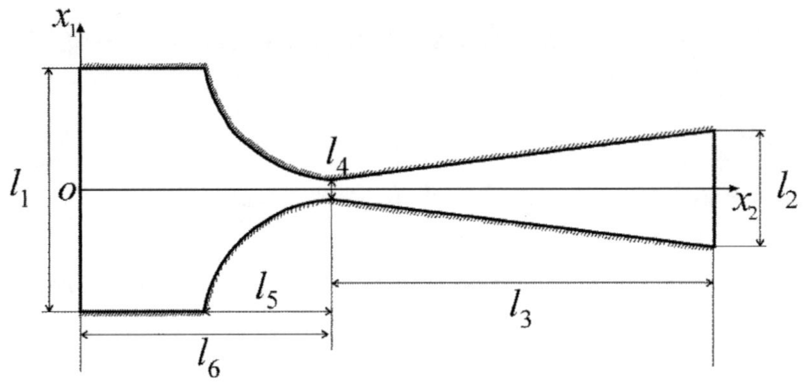

Figure 1. Computational domain of the Laval nozzle.

To simulate the motion of a mixture of liquid and vapor phases are used Favre-averaged Navier-Stokes equation for a homogeneous medium with a variable density, supplemented with the equation for the enthalpy transfer in the form [2-5]:

$$\frac{\partial \rho_m}{\partial t} + \frac{\partial(\rho_m u_j)}{\partial x_j} = 0, \tag{1}$$

$$\frac{\partial(\rho_m u_i)}{\partial t} + \frac{\partial(\rho_m u_i u_j)}{\partial x_j} = -\frac{\partial \hat{p}}{\partial x_i} + \frac{\partial}{\partial x_j}\left[(\mu_m + \mu_t)\left(\frac{\partial u_i}{\partial x_j} + \frac{\partial u_j}{\partial x_i} - \frac{2}{3}\frac{\partial u_k}{\partial x_k}\delta_{ij}\right)\right], \tag{2}$$

$$\frac{\partial}{\partial t}\left[\rho_m (h + f_V L)\right] + \frac{\partial}{\partial x_j}\left[\rho_m u_j (h + f_V L)\right] = \frac{\partial}{\partial x_j}\left[\left(\frac{\mu_m}{\Pr_L} + \frac{\mu_t}{\Pr_t}\right)\frac{\partial h}{\partial x_j}\right], \tag{3}$$

$$\frac{\partial \alpha_L}{\partial t} + \frac{\partial(\alpha_L u_j)}{\partial x_j} = \frac{1}{\rho_L}(\dot{m}^+ + \dot{m}^-) \tag{4}$$

There x_i - Cartesian coordinates, u_i - the components of the velocity vector, t -time, $\hat{p} = p + (2/3)\rho_m k$, p - pressure, k - kinetic energy of turbulent pulsations, $\rho_m = \alpha_L \rho_L + (1-\alpha_L)\rho_V$ - the density of the mixture, ρ_L - the density of the liquid, ρ_V - vapor density, α_L - the volume fraction of the liquid phase in the mixture, δ_{ij} - the Kronecker symbol, $\mu_m = \alpha_L \mu_L + (1-\alpha_L)\mu_V$ - dynamic coefficient viscosity of the mixture, μ_L, μ_V - the coefficients of dynamic viscosity of the liquid and vapor respectively, $h = c_p T$ -specific enthalpy, T - temperature, c_p - thermal capacity coefficient at constant pressure, f_V - is the mass fraction of vapor, defined by the relation $f_V = \rho_V (1-\alpha_L)/\rho_m$, L - latent heat of vaporization, $\Pr_L$, $\Pr_t$ - molecular and turbulent ($\Pr_t = 0.9$) Prandl number, $\dot{m}^-$, $\dot{m}^+$ - model source components, taking into account the process of vapor condensation and liquid evaporation. Assuming the presence of non-condensable gases in the liquid with a mass content f_g, in the paper [6] are represented equations for $\dot{m}^-$, $\dot{m}^+$ in the form of

IOP Conf. Series: Materials Science and Engineering **134** (2016) 012028 doi:10.1088/1757-899X/134/1/012028

$$\dot{m}^{+} = C_e \frac{\max(1{,}0,\sqrt{k})(1-f_V-f_g)}{\gamma} \rho_L \rho_V \sqrt{\left(\frac{2}{3}\frac{p_V(T)-p}{\rho_L}\right)}, \quad p<p_V(T), \tag{5}$$

$$\dot{m}^{-} = C_c \frac{\max(1{,}0,\sqrt{k}) f_V}{\gamma} \rho_L \rho_V \sqrt{\left(\frac{2}{3}\frac{p_V-p}{\rho_L}\right)}, \quad p \geq p_V(T) \tag{6}$$

In relations (5) - (6) $p_V(T)$ - saturation pressure of methane at the temperature T , γ - coefficient of surface tension of liquid methane. Empirical constants C_e, C_c that determine the rate of evaporation and speed of condensation and have values $C_e/\gamma = 1.665\cdot 10^3, C_c/\gamma = 1.225\cdot 10^3$. The coefficient of turbulent viscosity is calculated according to the formula of Kolmogorov-Prandl $\mu_t = C_\mu \rho_m k^2/\varepsilon$, where to determine the kinetic energy of turbulence k and rate of dissipation ε two-parametrical model of turbulence $k-\varepsilon$ is applied [7].

3. Results

For simulation of flow in a Laval nozzle a grid with 124000 finite volumes with a significant concentration in the area of throat constriction was used. In the calculation of turbulent flows special attention should be paid to mesh quality near the walls, in regions of large gradients of flow parameters. The flow in a Laval nozzle was considered for different values of inlet pressure in the nozzle p_0 = 0,57 MPa, 2.5 MPa. For low-pressure flow (p_0 =0,57 MPa) the influence of the degree of underheating on the hydrodynamic structure of the flow, while for high pressure (p_0 =2.5 MPa) counter-pressure parameter ε_a was investigated. In this paper the pressure variation along the nozzle axis for the flow at p_0 =0,57 MPa, $\varepsilon_a = 0{,}76$ and various degrees of underheating Δt_H of the incoming liquid (Δt_H =4,0^0, 12,0^0, 20,0^0 K) was investigated. In figure 2 shows the distribution of α_V in the computational domain at p_0 =0,57 MPa, $\varepsilon_a = 0{,}76$, Δt_H =4,0^0 K and p_0 =2,5 MPa, $\varepsilon_a = 0{,}36$, Δt_H =5,0^0 K. This figure illustrates that at the minimum section of the nozzle moving incompressible fluid, and the vaporization process develops immediately after the passage the nozzle throat. The characteristic "stretching" of the contour of variable α_V (volume content) in the central part indicates that the vapor phase are intensively formed in the boundary layer along the walls of the channel and then spreads to the nozzle axis. At some distance from the minimum section the liquid kernel disappears, and and bubble transitional structure will be transformed to the vapor-bubble flow.

In this paper to simulate the vaporization process a model with an additional transport equation of volume of phase with the source terms for modeling the speed of condensation and evaporation is used [3]. The model assumes that the rate of change of phase state is proportional to the characteristic speed, as which the relative velocity between the liquid and bubbles is selected. The relative velocity V_{ch} of the same order as the velocity of turbulent fluctuations, for which the estimate $V_{ch} \sim \sqrt{k}$ is fair. Comparing the calculated data on the distribution of the void fraction of vapor, is shown in figure. 2 it can be concluded that the structure of wet steam flow in the diffuser part of the nozzle for low-pressure and high-pressure flow is qualitatively similar, but have the some different features. The vapor phase in both cases initially is greater formed in the boundary layer near the walls of the channel and then will extend with the flow to the nozzle axis. The difference seen in the fact that the increase the velocity of bubbling flow in the Laval nozzle is accompanied by a decrease in the volumetric void fraction in the central part of the expanding portion of nozzle, but and in the near-wall regions it increases. For high-pressure flow rate of vaporization in the region near the wall increases due to the increase of turbulent kinetic energy near the walls and consequently the growth of the characteristic speed. Small nucleation particles that appeared near the walls, growing in size and are carried by flow to the central partion of the nozzle. Due to the increased flow rate, the time for the growth of bubbles decreases, and they more quickly propagate downstream, which leads that the volumetric void fraction near the axis of the nozzle is decreased.

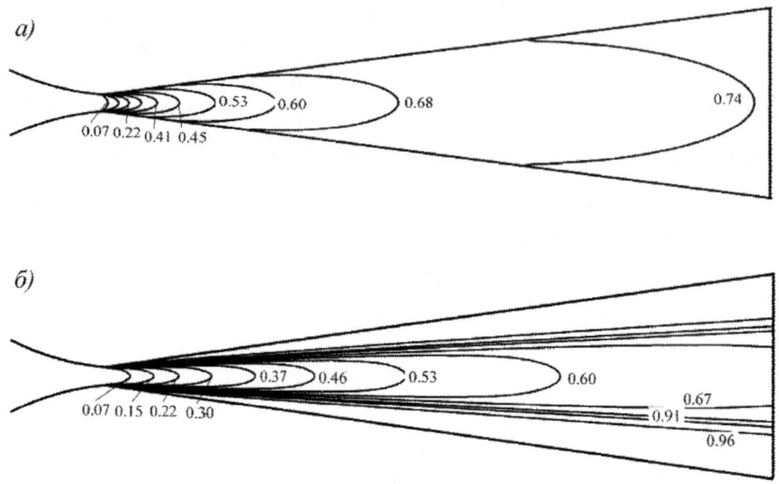

Figure 3. The contour of distributions of α_V for different operating parameters *a*) $p_0 = 0{,}57$ MPa, $\Delta t_H = 4.0^0\,\mathrm{K}$, $\varepsilon_a = 0{,}76$; *b*) $p_0 = 2{,}5\,\mathrm{MPa}$, $\Delta t_H = 5.0^0\,\mathrm{K}$, $\varepsilon_a = 0{,}36$

If at p_0 =0,57 MPa the value α_V which is gradually increasing along a channel axis on a cut of the channel equals α_V =0,75, at p_0 = 2,5 MPa it reaches only α_V = 0,65. The increasing the flow rate leads to a redistribution of the void fraction along the vertical sections of the diffuser part of the nozzle. The increase of counter-pressure at a significant initial underheating leads to condensation jumps of vapor-liquid flow in the expanding portion, in which there is a full condensation of a stream is occured.

The results of numerical simulation of flow boiling of methane in a cylindrical channel of variable cross section showed good agreement with the experimental data. Calculations showed that the hydrodynamic features of the flow in the Laval nozzles (length of liquid kernel, extent of bubble and vapor-liquid zones, the distribution of phases, physical variables of a flow in the field) depends on the geometrical shape of the channels and operating parameters (p_0 , Δt_H , ε_a).

Work is performed with financial support of the Ministry of Education and Science of the Russian Federation, the Agreement on a subsidy No. 14.577.21.0151 of 28.11.2014. The identifying number of the project RFMEFI57714X0151.

References

[1] Zusin B A, Baranov G A, Barilovich B A and Parfenova I H 1976 *Flushinag adiabatic flows* Atomizdat
[2] Nigmatulin R I 1978 *Fundamentals mechsnics of geretogenous media* Science
[3] Saifutdinov A I, Saifutdinova A A, Kashapov N F and Fadeev S A 2016 *J. Phys.: Conf. Ser.* **669** 012045
[4] Saifutdinov A I, Fadeev S A, Saifutdinova A A and Kashapov N F 2015 *JETP Lett.* **102** 637–642
[5] Gubaidullin D A , Morenko I V and Fedyaev V L 2015 *Russian Aeronautics (Iz VUZ)* **58** 351–354

[6] Singhal A K, Athavale M M and Jiang U 2002 *J. Fluids Eng ASME* **124** 617-624
[7] Kataoka I and Serizawa A 1989 *Int. J. Multiphase flows.* **15**(5). 843-855

IOP Conf. Series: Materials Science and Engineering **134** (2016) 012029 doi:10.1088/1757-899X/134/1/012029

Predicting structure micro-alloyed steel products for different purposes

V I Astaschenko, E A Zapadnova, N N Zapadnova and G F Mukhametzyanova

Kazan Federal University, Naberezhnye Chelny Institute, Pr. Mira 68/19, Naberezhnye Chelny, Republic of Tatarstan, Russia, 423812

e-mail: gulnara-ineka@mail.ru

Annotation. The composition of micro-alloyed steels for controlled blanks forging of the auto important parts and for the products operating at lower temperatures is predicted. It was found that the use of steels, micro-alloyed with vanadium and nitrogen, vanadium, aluminum and nitrogen are useful for controlled forging blanks of the auto critical parts and as a cold-resistant steels for use in Siberia and the Far North.

The alloying element in the steel is capable of changing its grain size, display (bring out) unwanted impurities (in particular, to react with the oxygen, arsenic or sulfur being there and, harmfully effecting a metal alloy), to raise the hardenability, heat resistance and other properties [1, 2].

Under microalloying we understand the small additives of elements or their compounds added to the composition of the alloy that enhance its properties through the local doping of grain boundaries and subgrains and the formation of strengthening phases with sharply limited solubility of the latter in the α-γ-solutions. The term "small" additives depends on the alloy composition and the solubility of the impurity in its main component [3].

Slightly different chemical elements are used in microalloying than in conventional doping. Those are reactive substances - zirconium, boron, tungsten, niobium, aluminum, and vanadium, titanium, and others. Typically Al, Ca, Mg, B, Ba, and N; and refractory metals (Zr, Ti, Nb, V), rare earths metals (Ce, La, Y et.) and their mixtures (mischmetal, ferrocerium) [4] are used for microalloying.

Billets of crucial parts in mechanical engineering are usually manufactured by hot plastic deformation (forging, stamping, rolling and etc.) as the deformable metal has a significant advantage in their properties, compared with a cast condition. Furthermore, for such items it is recommended to use steels with carbon content in the range 0.3-0.5%, alloyed with chromium, nickel, molybdenum and vanadium. However, these steels have a drawback - they are expensive and prone to overheating at temperatures of hot stamping or forging (1100-1200 ° C). There is a significant grain growth during heating the chromium-nickel steel. Steels alloyed with vanadium and titanium are less prone to overheating (Figure 1).

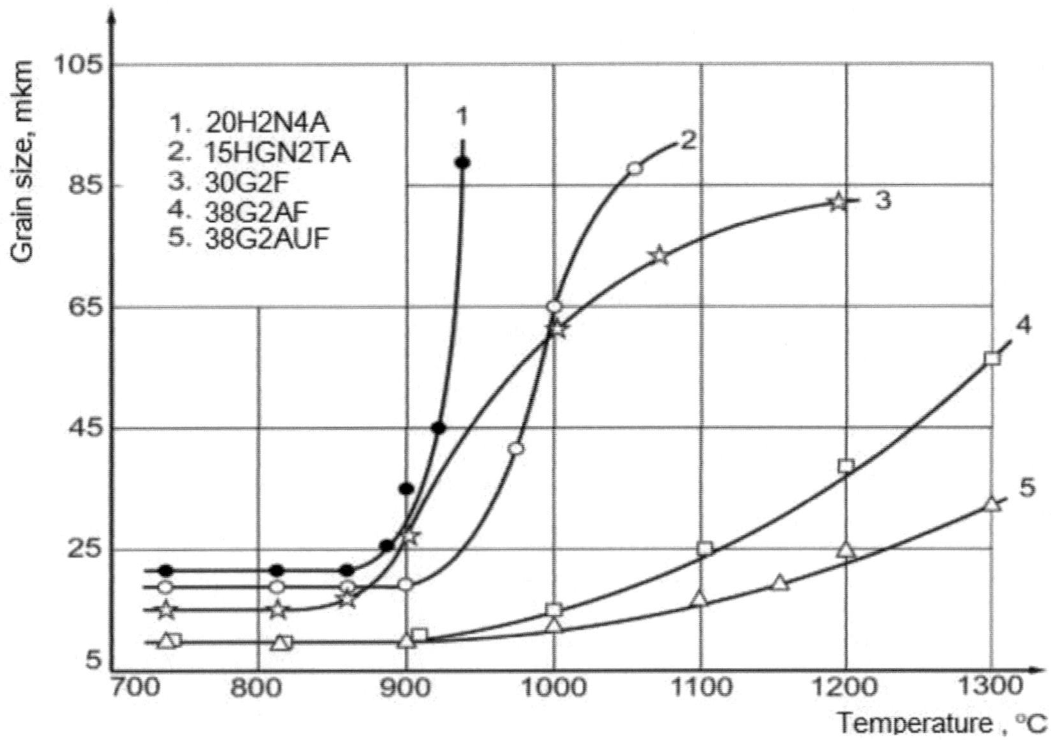

Fig. 1 - Temperature dependence of the grain growth in the steel.

One of the ways to solve these problems is steel microalloying with vanadium, aluminum, nitrogen [5, 6]. With all-inclusive introduction of these complex elements we have maximum technological and economic benefits. The slight grain growth occurs in such steels (Figure 1, curves 4 and 5) even at temperatures of hot plastic deformation beginning (1100-1200 ° C) . Steel high resistance to overheating is due to the formation of aluminum nitrides and nitrides of vanadium, which are located along the grains boundaries and hinder their growth.

The positive effect of microalloying with vanadium steel, aluminum and nitrogen can be realized in the production of billets by hot stamping by the use of the residual heat for the heat treatment of forgings. Fine structure maintaining during deformation and subsequent postdeformation cooling eliminates the need for steel re-heating above the critical temperature intended for recrystallization, and accordingly makes it possible to exclude the operation of austenization during heat treatment.

Steel structure formation can be controlled and a given structure and properties of the alloys can be formed during the implementation of controlled deformation processes and billets cooling. The positive role of micro-alloying is observed in low-alloy steels used in the construction industry, the oil and gas industry and the railways [7, 8]. Steel grades 18F 15GF are successfully used in the manufacture of building structures. The use of these steels in certain construction projects saved up to 20% of rolled steel and reduced the cost of rental structures by 8-15%. The use of steel grade 09G2 12HGFD in the manufacture of semi-carriages (gondolas) made it possible to obtain significant savings by reducing weight, increasing the reserve maintenance periods and increasing the semi-carriages (gondolas) reliability. For the operation in the North low carbon steels (0.3% C) are used doping them with different elements. For a long time only nickel steel 3, 6 and 9% of nickel and low carbon content or austenitic chromium-nickel steels were used for these temperature operating conditions. For a moderately low temperatures low carbon micro-alloyed steels with low nickel content up to 1.5% or even without it are recommended. Steel, micro-alloyed with vanadium and nitrogen (grade 15G2AF), has a high cold resistance with considerable strength that allows us to recommend it for the manufacture of large diameter pipes for trunk pipelines, working in the conditions of the North. Thus, the use of steel, micro-alloyed with vanadium and nitrogen, vanadium, aluminum and nitrogen are useful for controlled blanks forging of the auto critical parts and as cold-resistant steels for operating use in Siberia and the Far North.

References

[1] V G Shibakov, D L Pankratov, R V Shibakov 2014 Connection between machines durability during operation and the complex of physic-mechanical properties, formed during the production of critical parts *IOP Conference Series: Materials Science and Engineering* **69** 012040
[2] Goldstein Ya E, Mizin V G 1986 *The modification and microalloying of cast iron and steel* (Moscow: Metallurgiya) p 272
[3] Gavrilyuk V G and Efimenko S P 1990 *Nitrogen influence on the structure and properties of iron and perspective directions of development of high-nitrogen steels* (Kiev: Proceedings of the 1st All-Union Conference of High- nitrogen steels) pp 5-26
[4] Brown M P 1982 *Steel microalloying* (Kiev: Sciences Dumka) p 303
[5] Korchinsky M 2005 *Advanced metallic structural materials and a new role for microalloyed steels* Steel chapter 6 pp 124-130
[6] Fonstain N M 1990 *Vanadium microalloyed steels*, Steel chapter 6 pp 89-94
[7] Solntsev Yu P 2005 *Cold-resistant steel and alloys* (St. Petersburg: Khimizdat) p 480
[8] Odesski P D 2005 *Microalloying steels for the Northern and unique metal structures* Steel chapter 6 pp 116-123

Study of vibro-acoustic properties of composite materials based on polyurethane injection

S M Vakhitova, I F Shaehova, E I Zharin

Kazan Federal University, Naberezhnye Chelny Institute Pr. Mira 68/19, Naberezhnye Chelny, Republic of Tatarstan, Russia, 423812,

e-mail: vakhitova.svetlana@yandex.ru

Annotation. Modern standards of noise and vibration make the manufacturers of vehicles, machines and construction materials take into account the high demands in the design process. Often there is a need both in vibrations absorbing and in effective sound absorbers. Therefore there was a research on the concentrations of matrix components, fillers and modifiers impact on the absorbing and elastic properties of the polyurethane composites. The evaluation of sound-absorbing and elastic properties was in determination of sound absorption coefficient Cs and dynamic elasticity modulus Ed of developed composite materials. The aim of this work is to study changes in the properties of composite materials based on polyurethane matrix to identify compounds with the highest sound absorption and vibration absorbing properties simultaneously.

Urethane composition was used as matrix components, consisting of component A (mass fraction of hydroxyl groups, 4.4%); Component B (mass fraction of isocyanate groups, 21.4%). A polyurethane matrix selection is due to the high popularity of polyurethanes in modern industry due to a number of valuable properties, such as resistance to mechanical and chemical influences, a wide range of physical and mechanical properties and others.

Fillers and modifiers used:

1. Aerosil - colloidal silicon dioxide ($SiO2$), very light micronized powder with strong adsorption properties. "Aeros" is a trade name, coined by the German chemical company «Evonik Degussa AG». The technical name is pyrogenic silica.

2. Expanded clay - lightweight porous building material, produced by the burning of fusible clay.

3. Aluminum hydroxide - alumina compound with water. The white gelatinous substance poorly soluble in water, has amphoteric properties. Prepared by reaction of aluminum salts with aqueous solutions of alkalis, avoiding their excess.

The dried aluminum hydroxide is a white crystalline substance, insoluble in acids and alkalis. It is used as a flame retardant (suppressor burning) in plastics and other materials.

4. The organic silicon compound. There is a connection between the silicon and carbon atoms its

molecules. It is used for the production of lubricants, polymers, rubber, rubbers, silicone fluids and emulsions.

Directional control of the chemical structure of the polymers is a method of improving them. This process can be carried out by copolymerization of polymers with other monomers. Their introduction (injection) increases the mobility of macromolecular segments and increases the sound-absorbing properties.

In order to enhance the elastic and sound-absorbing properties the two-component polyurethane matrix was modified by the introduction of silicon organic compound (CO).

As a result of a chemical reaction of the polyethylenepolyamine with the polyisocyanate gel fraction was formed which subsequently serves as an additional source of dissipation of vibrational energy.

Figure 1. shows the effect of modifying agent (10 parts by weight) on dynamic modulus of elasticity (Ed) polyurethane CM with different concentrations of matrix components.

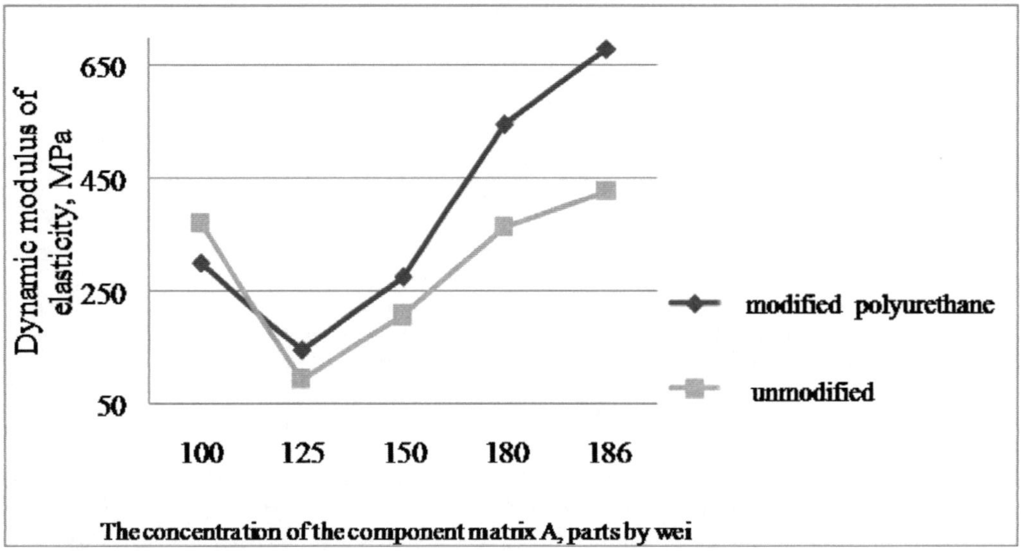

Figure 1. Effect of modifying agent (10 parts by weight) on Ed PU matrix CM with different concentrations of matrix components

As can be seen from the graph in Figure 1, the dynamic modulus of elasticity of the modified polyurethane CM varies extremely from a sharp decrease to an increase. The dynamic modulus of elasticity of nearly all the unmodified compositions is higher than of unmodified ones. This indicates that the modification has improved elastic properties of the PU matrix.

Table 1 shows that the sample with the concentration of the matrix components A: B is equal to 125: 100 and a modified CO with additive in volume 10 weight parts has a Cs greater than the unmodified composition. The sound absorption properties increase with the introduction of the complex reactive additive is connected with decreased velocity of wave propagation in the polymer. In addition, such a content of CO compound (10 parts by weight) helps to reduce the flammability of the material and expands the scope of its application.

Table 1. The effect of modifier (10 parts by weight) on the sound absorption coefficient of matrix polyurethane CM

modifier	component A	component Б	sound absorption coefficient Cs		
			630 Hz	1000 Hz	1600 Hz

	125	100	0,15	0,30	0,06
CO 10 parts by weight	125	100	0,19	0,35	0,08

It is known that water is widespread intumescent agent of PU. The amount of water required for active gassing of polyurethane matrix is 1%. The porous structure effectively absorbs sound waves and dampens vibration.

The interaction of a certain ratio of matrix components with the modifying additive forms PU foam. Foamed CM shows the socyanate component sensitivity to moisture, which is initially in uretanoforming polyester and is absorbed from the environment.

The analysis of data obtained as a result of the study showed the feasibility of modifying the PU matrix by one-component modifying additive at 10 parts by weight and the introduction of 1% water to obtain the CM with higher sound-absorbing and elastic properties compared with the PU matrix composites. All subsequent CM filled compositions are based on PU composition modified by 10 parts by weight of the compound CO and 1% water.

Table 2 shows the compositions of CM samples with different ratios of fillers and the study results of their sound absorption coefficient and the dynamic modulus of elasticity.

№ samples	Components ratio of modified matrix, parts by weight		Filler and its content (parts by weight)	Sound absorption coefficient Cs, in Hz			Dynamic modulus of elasticity Ed Mpa
	A(GSS)	Б(PFP)		630	1000	1600	
1	186	100	–	0,07	0,07	0,05	426
2	175	100	–	0,04	0,11	0,04	364
3	150	100	–	0,06	0,11	0,01	207
4	125	100	–	0,15	0,3	0,06	93
5	100	100	–	0,08	0,14	0,03	371
6	125	100	Aerosil ; 0,1	0,35	0,12	0,08	761
7	125	100	Aerosil; 0,2	0,43	0,37	0,26	165
8	125	100	Aerosil; 0,3	0,48	0,57	0,46	95
9	125	100	Expanded clay; 0,1	0,25	0,27	0,017	328
10	125	100	Expanded clay; 0,3	0,42	0,28	0,22	110
11	125	100	Expanded clay; 0,4	0,36	0,22	0,17	41
12	125	100	Aluminum hydroxide, 0,2	0,25	0,21	0,2	338
13	125	100	Aluminum hydroxide, 0,3	0,33	0,75	0,05	326

Considering the results of samples tests of 1-5 we see that with increasing concentration of the matrix component A there is a growth dynamic modulus of elasticity, which is connected with an increase in the degree of PU monomer crosslinking.

The study of unfilled modified PU compositions with different ratio of components A: B shows that the modified matrix composition in a ratio of 125: 100 has the greatest Cs. The same composition showed the smallest elastic modulus Ed, and thus the best vibration-absorbing properties among the rest of the samples.

The subsequent filling of the composition with aerosil, expanded clay and aluminum hydroxide increased sound absorption and reduced dynamic modulus of elasticity. It should be noted that with an increase in the filler volume content, the sound-absorbing and vibration-absorbing properties of the samples are increasing at the same time. But the growth is uneven.

The material with the best sound absorption (Ed = 41 MPa), sample №11, but the Cs in the entire investigated audio range did not exceed the value of 0.36. This means that being a good vibration-absorbing material, this material is not enough an efficient sound absorber.

The material with the best sound absorption properties is sample № 13. Maximum Cs occurs at a frequency of 1000 Hz and is 0.75. It should be noted that at other frequencies Cs of this sample shows much lower values. The storage modulus is 326 MPa, that indicates the lowest vibration damping capability.

The best of the investigated CM filled on the basis of the modified PU matrix is sample № 8, since in the entire frequency range from 800 Hz to 1600 it showed a stable performance of Cs, increasing with the filler weight increase to a value of 0.57. The modulus of elasticity of sample number 8 is 95 MPa, that means good vibration-absorbing ability.

The analysis of the data obtained showed the advisability of PU matrix modification by one-component modifying additive in a concentration of 10 parts by weight and its subsequent filling for vibro-acoustic CM obtain.

Bibliography:

[1] Solomatov V I Cherkasov V D and Fomin N E 2001 *Vibration-absorbing composite* – Saransk: Pub. Mordov. University Press 96 p

[2] Cherkasov V D 1995 *Building composites with improved vibration-absorbing properties* Moscow: Synopsis. Dis. Doctor. tehn. Science Cherkasov V I 40 p

[3] Zharin D E 2006 *Scientific fundamentals of vibration-absorbing construction materials* Kazan: Synopsis. Dis. Doctor. tehn. Science Zharin D E - 45 p

[4] Novikov V U 1995 *Polymeric materials for the construction: a handbook* (Moscow: Higher. Sch. - 448 p

[5] Barth V E, Sanin G S and Shevchuk S A 1985 *Polymer concrete usage in machine tooling* Moscow: VNIITEMR 40 p

[6] Bazhenov Y M 1970 *The concrete under dynamic loading* Moscow: Stroyizdat 272 p

[7] Gumerov I F, Shafigullin L N, Vakhitova S M and Shaekhova I F 2014 *Noise absorbing composite materials applied in domestic trucks* IOP Conference Series: Materials Science and Engineering **69** 012011

[8] Galyautdinov R T, Kasparov N F and Luchkin G S 2002 *Technology of sputtering high-reflection coatings on ABS-plastic articles* Inzhenerno-FizicheskiiZhurnal **75** 170–173

IOP Conf. Series: Materials Science and Engineering **134** (2016) 012031 doi:10.1088/1757-899X/134/1/012031

Influence of multilayer coatings on the operational stability of molds for injection molding

D L Pankratov, R. V. Gavariev[2], K. N. Gavarieva[1]

[1] Kazan Federal University, Naberezhnye Chelny Institute
Pr. Mira 68/19, Naberezhnye Chelny, Republic of Tatarstan, Russia, 423812
[2] Kazan National Research Technical University. AN Tupolev-KAI Kazan, st. Karl Marx, 10, Russia, Republic of Tatarstan, 42011,

e-mail: Soldatkinakn@mail.ru

Annotation. The article examines the main factors affecting the operational stability of molds for injection molding of non-ferrous metals and zinc alloys in particular, and also provides a specific method to increase the index of resistance of molds by coating the molding surface by cathode- ion bombardment. It describes the effect of the proposed coatings on the parameters of pressure, tension and heat in a mold.

1. Introduction

Operational stability tooling increase is a very urgent problem of all foundries and engineering in general. Especially effective solution to this problem is in mass production processes of finished products, which include injection molding. The number of products manufactured in this manner is measured, usually in tens of thousands. The increased operational stability of molds for injection molding, even a small percentage, calculated on the number of manufactured products will have a significant positive result.

There are many factors that have a negative effect on the operational stability of molds, but the most significant are cyclically repeated thermal power and influence on the shaping surface of the mold.

Modern theoretical recommendations can improve certain parameters affecting the operational resistance. They are: the proposals on the use of a special grade of steel as the material of the mold [1, 2], the special methods for calculating the operational stability of molds [3], the use of different lubricants and coatings [4]. However, most of them are related to the problem of stability in injection molding in a narrow range of process parameters regimented or using only aluminum alloys, which is not applicable to zinc alloys, owing to significant differences in melting point temperatures: 660 ° C and 380-480 ° C in aluminum and zinc alloys, respectively, and also because of the different chemical and physical activity.

IOP Conf. Series: Materials Science and Engineering **134** (2016) 012031 doi:10.1088/1757-899X/134/1/012031

There are some solutions to improve the operational stability of the molds, almost regardless of the initial conditions, such as the use of molybdenum tooling, but from the standpoint of economic efficiency such decisions are hardly applicable in today's workplace.

As a rule, nitriding is used in practice. This type of thermochemical treatment has several advantages: increased surface hardness, wear resistance, corrosion resistance, however, a significant disadvantage of this treatment is the brittleness increase of both the surface layer, and the material of the mold, which leads to the full swing cracks and as a consequence, to a short period of work. That is, this method is aimed primarily for aluminum alloys, which molds for casting can withstand up to 50,000 pressings, while for zinc alloys according to the State standards GOST 19946-74, it's necessary to ensure a minimum rate of 150000 pressings.

2. The main part

Thus, the problem of improving the operational stability of molds for injection molding of zinc alloys is relevant and requires further research. The solution to this problem can be multilayer coatings deposited by cathode ion bombardment [5,6,7]. These coatings have a number of advantages, which distinguish them from other possible solutions. Among them, the use of substantially any refractory or heat-resistant metal as a base coating, the possibility of applying several layers having specific, predetermined properties, hardness is also the possibility of layers alternation to inhibit the emergence and growth of cracks. In addition, each new layer does not affect the previous ones.

Thus, a possible solution to the operational durability of molds can be a multilayer coating consisting of a carbonitride molybdenum (MoCN) lower layer 2 microns thick for adhesion of the coating to the mold,metal surface and then on top of the lower layer an intermediate layer of 3 microns thick of nitride titanium (TiN) to provide high hardness of the coating, followed by a top layer with thickness of 2 microns of molybdenum nitride (MoN) for abnormally low coefficient of friction.

Using this coating without determining stresses acting on the molding surface of the mold, and without making the heat balance between the mold and casting is unacceptable. The stresses caused by power factors are completely dependent on the magnitude of the pressure acting on the wall of the shaping mold. An important point is the configuration of the surface contacting with the molten metal. In general, the pressure on the wall can be calculated according to the laws of hydraulics [8]:

$$p_{уд} = \gamma_{ж} \frac{\omega_{в}^{2}}{g}, \qquad (1)$$

wherein $p_{уд}$ - the pressure on the wall, $\omega_{в}$ - the speed of the molten metal inlet, $\gamma_{ж}$ - the volumetric weight of the molten metal, g - acceleration of gravity.

However, if the surface is different from a straight wall, it is necessary to introduce correction factors, so for a concave surface the pressure is doubled:

$$p_{уд} = 2\gamma_{ж} \frac{\omega_{в}^{2}}{g}. \qquad (2)$$

In the case of the concave surface, the pressure will be determined as follows:

$$p_{уд} = \gamma_{ж} \frac{\omega_{в}^{2}}{g}(1-\cos\varphi) \qquad (3)$$

where φ - the angle between the initial direction of flow and its direction after a meeting with the barrier.

Despite the well-known formulas for determining the pressure on concrete surfaces, the definition of specific pressure in a particular mold is a difficult task, since molding surface is a combination of different surfaces smoothly transitioning into each other. Therefore, the confirmation of the positive impact of multilayer coatings applied by CIB method can be a series of experiments in which the pressure and the stress are measured on distinct surfaces with the use of the proposed coating on surfaces with nitriding.

Multilayer coating based on molybdenum tests showed the following results: a slight reduction of pressure on the straight and curved surface does not exceed 1% of the pressure exerted on similar nitrided surfaces. However, the proposed coverage on the concave surface indicated the decline in specific pressure up to 5%, which is especially important when you consider that the most loaded sections of the forming surfaces of molds are precisely concave surfaces, such as hollow chamfer, fillet radius.

These results can be explained by the fact that the coating deposited (inflicted) by CIB significantly reduce the coefficient of friction between the molding surface and the molten metal, due to the physical nature of the molecular building coverage on the molding surface the purity is improved, but also when casting the physical interaction of the mold with the molten metal is decreased.

Thus, a coating deposited (inflicted) by CIB positively affect the decline in pressure and stress on the molding surface of the mold.

At the same time without making the heat balance between the mold and the product of casting it is impossible to make any conclusions about the effectiveness of the proposed coating.

The following papers [8,9,10,11] are devoted to research of shaping processes on hidher temperatures. To determine the thermal balance of the mold, it is generally necessary to determine its two components, one of them Q_n - the input heat and Q_o - the waste heat. The mutual equality of these quantities defines the thermal balance of the mold and the casting; otherwise it demands preliminary heating or cooling the mold. Input heat can be calculated according to the known procedure [12]:

$$Q_n = Q_{nep} + Q_{кр} + Q_{om} + Q_{mp}, \quad (4)$$

where $Q_{êð}$ - the heat of crystallization;

$Q_{êð}$ - the heat of solid casting;

$Q_{кр}$ - frictional heat;

Q_{nep} - overheating of the metal heat.

The amount of the last two components does not exceed 12% of the total value of the amount of input heat.

Equally challenging is the calculation of waste heat, as there are many gaps in the mold. Despite the complexity of the practical point of view, waste heat from the die is defined by two main components. One of them Q_T - this heat is rejected from the closed mold through the side surface, the other Q_P - a waste heat from the forming surface in the open state of the mold. Thus, the waste heat can be calculated as follows:

$$Q_o = Q_T + Q_P \ ; (5)$$

According to the coatings deposited by cathode ion bombardment experimental studies [13], the following results were obtained: the composition of the coating varying did not produce a significant decrease in indicators of heat input and the heat released into the environment and the change in their total value amounted to not more than 1%. This is due to a sufficiently small thickness of the proposed coating not more than 10 microns. However, reducing the coefficient of friction between the molding surface and the flow of molten metal provides the reduction in input heat to 3%, which is 10-12 ° C for zinc alloys.

3. Conclusion

Thus, the coating consisting of the carbonitride molybdenum (MoCN) lower layer of 2 microns thick for adhesion of the coating to the metal surface of the mold, and then of the intermediate layer of 3 microns thick of titanium nitride (TiN) on top of the lower layer to provide high hardness of the coating, and also the molybdenum nitride (MoN) top layer 2 microns thick for abnormally low

IOP Conf. Series: Materials Science and Engineering **134** (2016) 012031 doi:10.1088/1757-899X/134/1/012031

coefficient of friction, provides a decrease in the parameters of pressure and stress on the working surfaces, particularly of a concave shape, as well as the input heat reducing by receiving an abnormally low coefficient of friction between the molding surface and the flow of molten metal, that ultimately will improve the operational stability of molds for zinc alloys casting under pressure.

Bibliography:

[1] Nemtyrev O V 2005 *Development and Research of High-Resistant Chromium Steel Casting Molds for Injection Molding of Aluminum Alloys* (Rybinsk: Synopsis Dis. ... Cand. Tehn. Sciences) p 16

[2] Denisov P Yu 2005 *Interaction of Alloys of Aluminum Materials Molds and Their Resistance Increase with Functional Coatings* (Tyumen: Synopsis Dis. ... Cand. Tehn. Sciences) p 16

[3] Berezin E T 2002 *Increased Operational Stability of Molds Injection Molding Based on the Analysis of Heat-Stressed Condition and Process Modeling Thermal Fatigue Fracture* (Rybinsk: Synopsis Dis. ... Cand. Tehn. Sciences) p 28

[4] Sobolnitsky G D, Kononov V A, Tupchy Yu P, Sokolovsky A G and Smirnov Yu N 1991 *A Method for Producing a Thermal Barrier Coating on a Metal Mold for Casting Parts from Aluminum Alloys* (RF patent number 1,678,508, B22C23 / 02. 4773314/02 stated 12/25/89; publ. 1991.23.09.) Bull. 1 p 3

[5] Savin I A 2012 *The Study Wear-Resistant Coatings Characteristics Applied to Cutting Tools of Complex Shape by Ion Bombardment Cathode* (Blank production in mechanical engineering) pp 41-44

[6] Nozhenkov M V 2013 *The Structure and Properties of Composite Ion-Plasma Coatings* (Moscow: friction and lubrication in machinery Publisher: Scientific and technical publishing house "Engineering") 6 pp 3-12

[7] Goncharov M V and Goncharov V S 2012 *Ion-Plasma Layered Composite Coatings with Special Properties* (Togliatti: science - industry and service publishing: Volga Region State University of Service) 6-3 pp 69-76

[8] Pankratov D L , Shibakov V G , Valiev A M and Valieva R F 2015 *Technology of Manufacturing of Multiple Core Stamp Tools by Plastic Deformation* (IOP Conference Series: Materials Science and Engineering) **86**

[9] Safronov N N and Kharisov L R 2015 *The Use of SHS-Process Slag for the Preparation of Foundry Sand. Journal of Physics: Conference Series* (IOP Conference Series: Materials Science and Engineering) **86**

[10] Shibakov V G , Pankratov D L , Andreev A P and Andreeva S.I. 2015 *Prediction of Samples Failure During Severe Plastic Deformation by Multiple Extrusion* (IOP Conference Series: Materials Science and Engineering) **86**

[11] Pankratov D L and Kashapova L R 2015 *Methodology Development for the Sustainability Process Assessment of Sheet Metal Forming of Complex-Shaped Products* (IOP Conference Series: Materials Science and Engineering) **86**

[12] Goryunov I.I. 1974 *Molds for Injection Molding* (Leningrad: Engineering,) p 255

[13] Gavariev R V, Savin I A and Leushin I O 2014 *Improving the Efficiency of the Injection Molding Zinc Alloys by Optimizing the Thermal Balance of Molds* (Moscow: foundry) 7 pp 27-30

[14] KashapovL N, KashapovN F and KashapovR N 2013 *Investigation of the influence of plasma-electrolytic processing on the surface of austenitic chromium-nickel steels.* IOP Conference Series: Materials Science and Engineering *J. Phys.: Conf. Ser.***479** 012003

IOP Conf. Series: Materials Science and Engineering **134** (2016) 012032 doi:10.1088/1757-899X/134/1/012032

An expert system for ensuring the reliability of the technological process of cold sheet metal forming

L R Kashapova, D L Pankratov and P P Utyaganov

Kazan Federal University, Naberezhnye Chelny Institute
Pr. Mira 68/19, Naberezhnye Chelny, Republic of Tatarstan, Russia, 423812

e-mail: pankratovdl@gmail.com

Abstract. In order to exclude periodic defects in the parts manufacturing obtained by cold sheet metal forming a method of automated estimation of technological process reliability was developed. The technique is based on the analysis of reliability factors: detail construction, material, mechanical and physical requirements; hardware settings, tool characteristics, etc. In the work the expert system is presented based on a statistical accumulation of the knowledge of the operator (technologist) and decisions of control algorithms.

Cold stamping allows obtaining parts with complex shapes, manufacturing of which by other treatment methods are either impossible or difficult: body parts, frames, amplifying elements of car body, airplanes, etc. [1] However, the periodic appearance of defective goods in the production of the phenomenon is not rare. Such defects as dents, creases and tear sheet (or cracks) appear.

In modern market conditions: import substitution, saving material and labor resources, meeting the requirements of quality management, one of the urgent tasks is the problem of ensuring the reliability of technological processes.

In most cases, the cause of periodic defective goods is a negative result of the cumulative impact factors of technological process, which are permissible for a given material and technology ranges.

Therefore, at the stage of development and realization of technological processes of sheet metal forming it is necessary to establish, monitor, and manage their design and technological parameters to ensure their desired consistency.

The paper presents a method of automated estimation of reliability of technological processes, at the design stage allows to identify and correct these critical parameters.

The technique is the following: automated analysis of safety factors –detail design, material from which it is made, mechanical and physical requirements; the equipment parameters, which will be stamping, tool characteristics, etc. The list of factors and their labels are presented in [2,3,4].

The indicators of the chemical composition are denoted by coefficients reflecting the quantitative range of the content of impurities in steel. The coefficients on the content of carbon K_{M1}, sulfur K_{M2}, phosphorus K_{M3}, silicon K_{M4}.

Criteria for characterizing the mechanical properties of the material, presented in the form of private coefficients are the coefficients of the yield strength K_{M5}, tensile strength K_{M6}, relative elongation K_{M7}, relative narrowing K_{M8}, anisotropy K_{M9}.

Indicators that account for the punching technology, are represented by the coefficients of friction between the surface of the workpiece and the working parts of the tool K_{T1}, the permissible deviation

IOP Conf. Series: Materials Science and Engineering **134** (2016) 012032 doi:10.1088/1757-899X/134/1/012032

for the thickness of the initial sheet K_{T2}, temperature K_{T3}, permissible deviations on dimensions of the original sheet K_{T4} and limit the elongation factor K_{T5}.

Characteristics of the instrument are indicated by the following coefficients: coefficient reflecting the hardness of the material from which the working parts of the instrument are made $K_{Ш1}$, a coefficient reflecting the roughness value of the surface of the working parts of the instrument $K_{Ш2}$, a coefficient taking into account the wear of the working parts of the instrument $K_{Ш3}$, a coefficient taking into account the misalignment of the upper and lower parts of the stamp $K_{Ш4}$, the coefficient taking into account the irregularity of the clearance between the working parts of the stamp $K_{Ш5}$.

An important element affecting the reliability of the technological process of sheet metal forming is forging equipment. Its characteristics are presented by coefficients reflecting deviations of a nominal force $K_{Об.1}$, stroke $K_{Об.2}$ and shut height of the press $K_{Об.3}$.

Information on listed indicators (K_{Mi}, K_{Tj}, $K_{Об.k}$, $K_{Шg}$) is represented graphically in the reliability profile on figure 1, where conditions of three comparable technological processes are conventionally displayed: * reliable technological process; • the process with the periodic occurrence of defective goods; □ – process with systematic defective goods.

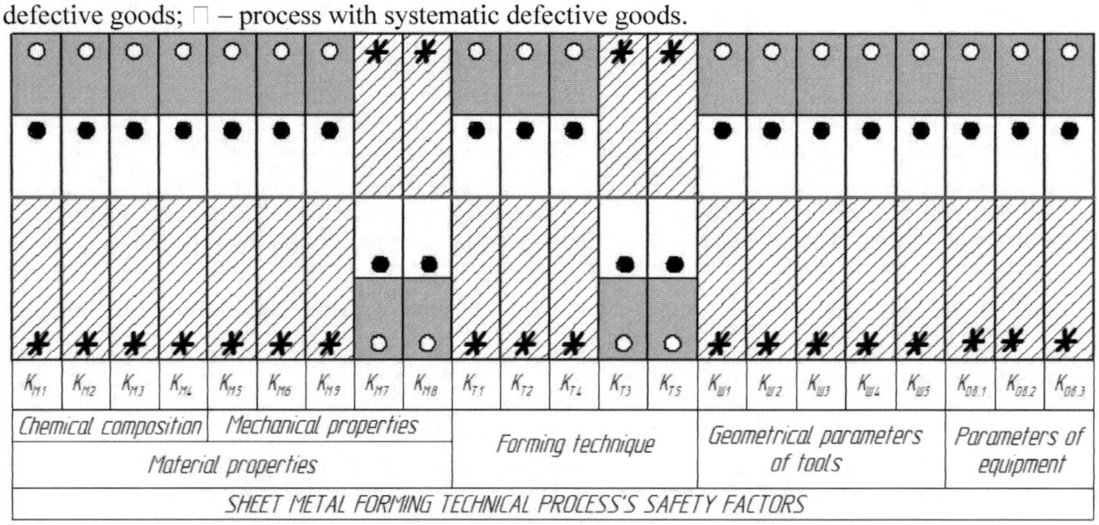

Figure 1 – Profile of indicators of reliability of technological process

As for some indicators the implementation of the process without occurrence of defective goods will contribute to their minimum value and some maximum, the profile of indicators of reliability of technological process of sheet metal forming is divided into three areas:

1) hatched – area in which the values of reliability indicators of sustainable technological process are desirable to be;

2) white, no fill – area in which reliability indicators for sustainable technological process are allowed to appear, starting from the middle line until the middle of the yellow area;

3) grey, with fill – area in which the values of indicators of reliability are undesirable because the probability of obtaining a defect-free process is far below (or equal to 0) than when injected into the area of green or yellow. [5]

Reliability profile is built based on the "radar" or "profiles" methods, which allows to combine different process parameters into a single dimensionless rate – integral coefficient of reliability and to objectively evaluate and compare processes. [5]

Expert system was developed based on a statistical accumulation of the knowledge of the operator (technologist) and decisions of control algorithms. The scheme of the expert system is presented in figure 2.

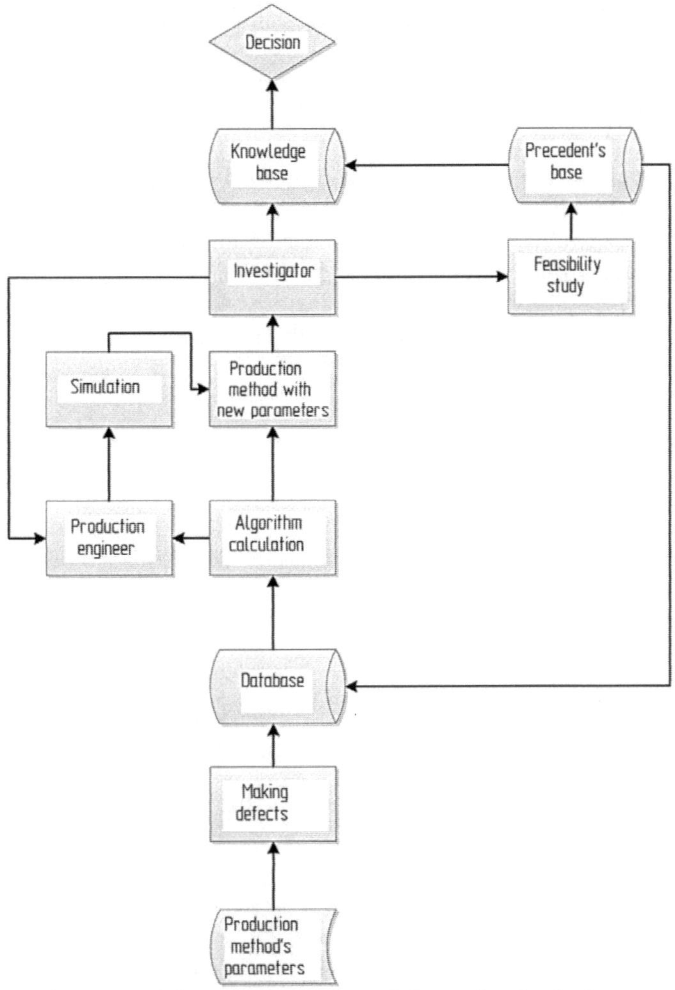

Figure 2 – Schematic diagram of the expert system of the technological process reliability of sheet metal forming

The core of the system is the knowledge base, which stores information about the dependencies between different safety factors. Knowledge is stored in the form of mathematical dependences and technological expertise, formalization of this knowledge is in the form of rules. The knowledge base is initially formed at the expense of formalized knowledge of experts. Further replenishment takes place either by the expert or on the basis of precedents.

The system operates on the base of the operation type and the corresponding parameters of the technological process (material grade, gauge, type of workpiece, lubrication, tool options, etc.) that are injected at the first stage.

Then goes the selection of defective goods which occasionally occurs in the stamping process. In the database in accordance with the type of defective goods the corresponding algorithm for calculation of technological parameters is chosen. The next step consists of calculation of technological parameters for selected algorithm, in the case of a successful calculation, the output is a technological process with the new settings. In case of impossibility of parameters calculation due to the algorithm, the technologist manually develops new process and approves it by means of simulation (for example, in the program AUTOFORM, PAMSTAMP, etc.).

In the "Expert" technological process with estimated parameters is checked by a specialist, who eventually accepts one of decisions: to send the process into the following stages: assessment of feasibility; return to the technologist for completion, send to the knowledge base, as a technology process with options that allow to obtain a defect-free stamping.

Feasibility assessment is an assessment of the reality and economic feasibility of implementation in specific production conditions.

After assessing feasibility, the data is entered into the database of precedents, from which it is transferred to the knowledge base and database.

The development of an intelligent system summarizing knowledge in the field of cold sheet metal forming, allows the calculation of rational parameters of technological processes excluding the possibility of periodic defective goods.

Thus, by adjusting the value of one or more reliability indicators, effect of interdependence "adjust" other indicators, can be explicitly or indirectly, to control the technological process based on intelligent systems. [6]

References:

[1] Romanovskiy V P 1979 *Handbook of cold stamping* Spravochnik po holodnoy shtampovke Spb.: Mashinostroenie 1979 520 p

[2] Kashapova L R, Pankratov D L and Shibakov V G 2014 *A technique for the automated estimation of reliability of technological process of sheet metal forming at the stage of its design* Fundamental research 8 (part 7) pp 1533-38

[3] Pankratov D L 2015 Methodology development for the sustainability process assessment of sheet metal forming of complex-shaped products *IOP Conference Series: Materials Science and Engineering* **8626** 012003

[4] L R Kashapova D L Pankratov A A Bilyalova 2014 Imrovement of the cold forming technology of the parts such as longeron *IOP Conference Series: Materials Science and Engineering* 69 012035

[5] Fashiev H A and Valeev I D 2007 *Differential method for assessing the quality of the vehicles* [Raznostnyiy metod otsenki kachestva avtomobiley] The automotive industry 11 pp 3-7

[6] Kashapova L R, Pankratov D L, Shibakov V G i Vinogradov A I 2014 *Stability and reliability procurement of the technological process of sheet stamping parts of complex shape* [Obespechenie stabilnosti i nadezhnosti tehnologicheskogo protsessa listovoy shtampovki detaley slozhnoy formyi] Materials of the international scientifically-practical conference "Innovative mechanical engineering technologies, equipment and materials-2014" pp 53-57

IOP Conf. Series: Materials Science and Engineering **134** (2016) 012033 doi:10.1088/1757-899X/134/1/012033

The erosion resistance of tool alloys in foundry melt the Zamak 4 – 1

GF Muhametzyanova, M S Kolesnikov and I R Muhametzyanov

Kazan Federal University, Naberezhnye Chelny Institute Pr. Mira 68/19, Naberezhnye Chelny, Republic of Tatarstan, Russia, 423812

e-mail: gulnara-ineka@mail.ru

Abstract.
The paper considers the resistance against erosion dissolution in the melt of foundry Zamak 4 – 1 die steels used for press machine parts manufacturing for injection molding, and hard alloys system WC – Co. It is established that the solubility in the melt Zamak – 4 – 1 steel of 4H5MFS and DI – 22 are promising for the parts fabrication of metal-wire casting machines of CLT and IDRA types. A significant reserve to increase the resistance of metal wires is the use of cast steel, as well as in electroslag and electro-beam remelting options. Metal-ceramic alloy doped with chromium VK25H may be recommended for reinforcement of heavily loaded parts of the press-nodes of hot casting machines under pressure.

The main cause of failure of parts of metal-wire and press-nodes components (mouthpieces, adapters, diffusers, liners, press-pistons, compression rings) of hot casting machines under pressure types CLT and IDRA experiencing hydrodynamic effects of zinc alloys, is erosive wear (feathering) of working surfaces, due to high physic-chemical activity of zinc interaction of melts with alloys based on iron, accompanied by the dissolution of iron and formation in the diffusion layer of aluminum and zinc intermetallic compounds of the Fe_2Al_5, Fe_2Zn_{13}, Fe_2Zn_7, etc., as well as solid solutions of η and α – phases [1, 2, 3, 4, 5].

This paper presents a comparative study of resistance to dissolution (erosive wear) in a zinc melt of foundry, Zamak 4 – 1 die steels: 4H5MFS, 3H2V8F, 5H4SCV4MF, 4H3VMF, 4H4VMFS (DI – 22), 5H3V3MFS (DI – 23), DI – 22 in electroslag (DI – 22 ESR) and electro-beam remelting (DI – 22EBR) variants, DI – 22 in a molten state, and sintered alloy of WC – Co with different content of cobalt: VK10, VK20, VK25, VK30; the size of carbide particles:(≈ 10 μm) – VK10S, VK20S; (≈ 30MKM) – VK20K; and doped(≈ 1%) with chromium VK25H, and molybdenum, VK25MO.

All-steel were thermally processed: after quenching from temperatures 1313 – 1393K and tempering 853 – 993K on the hardness of 44 – 46 HRC.

Hard alloys after sintering had hardness in the range 1750HV for VK10 and 1570HV for alloy VK30.

Material tests on the solubility in the melts were carried out in conditions of continuous change of the contacting fluid volumes and its flow impact on the sample typical for operational details loading

IOP Conf. Series: Materials Science and Engineering **134** (2016) 012033 doi:10.1088/1757-899X/134/1/012033

of metal-wire. Main dimensions: length, l = 30mm; diameter, d = 4mm; taper, ⊿ 15^0; the radius of curvature of the tops of the stake-owls, r = 0.02 mm. Investigated surface-cone of samples was processed to a purity of R_a = 0,8 – 0,64 microns and was controlled by the profilograph – profilometer – CALIBR. Model mode samples load: amplitude, 10.0 mm; the frequency, 100c^{-1}.The maximum speed of the movement of the studied plots in the melt is 2.0 m/sec. As criterion of solubility evaluation a change in the length of the sample, Δl, mm was adopted.

The study of kinetics solubility of steels and hard alloys in the melt Zamak – 4 – 1 was done under the duration of the interaction in the range of 12 – 90 min. at a constant temperature of melt, equal 743K (Fig. 1).

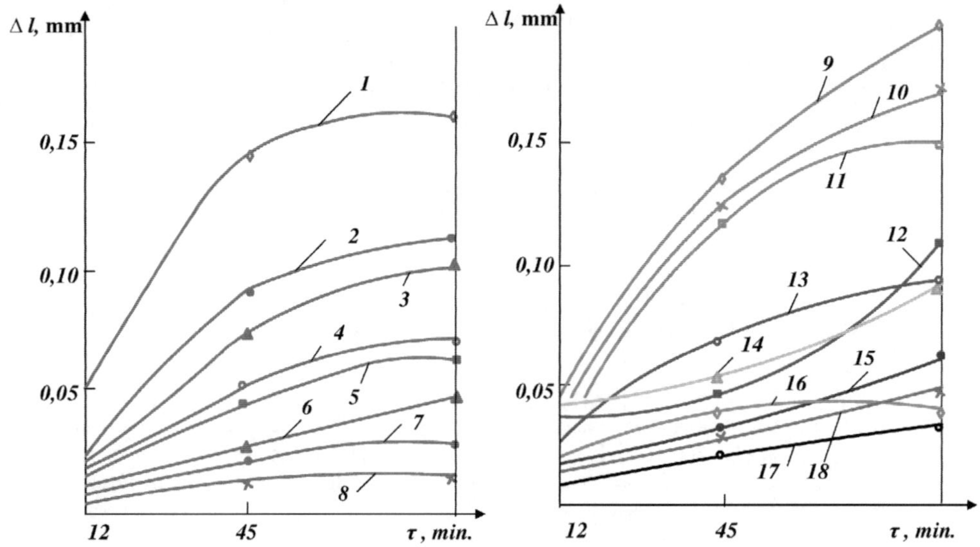

Fig.1 – Influence of duration of interaction (τ, min) on the solubility of (Δl, mm) tool materials in the melt Zamak – 4 – 1: 1 – 3H2V8F; 2 – 4X3VMF; 3 – 5H4SV4MF; 4 DI 23; 5 DI 22; 6 – 4H5MFS; 7 DI 22ESR; 8 DI 22ЭLP; 9 – VK30; 10 – VK25; 11 – VK20; 12 – VK10S; 13 – VK20S; 14 – VK20K; 15 – VK25M; 16 – VK10; 17 – VK25H; 18 – DI 22L.

Analysis of these data has confirmed that the solubility of the investigated materials in the melt Zamak – 4 – 1 varies considerably. For example, when the temperature is 743K and τ = 45 min, it changes from Δl = 0.012 mm (steel DI – 22 EBR) to Δl = 0.14 mm (steel 3H2V8F, curves 8 and 1 Fig. 1). A positive effect on the increase in the resistance of the solubility investigated steels has their alloying with chromium in combination with molybdenum and silicon (curves 3, 4, 5, 6, Fig. 1).

Reserve for increasing the resistance against dissolution is manifested in the use of material metal purity, i.e. after the ESR and EBR, as well as in molten state (DI – 22L, curves 7, 8, 18, Fig. 1).

Study of the effect of temperature on the solubility of materials in the melt Zamak – 4 – 1 was carried out in the range 723 – 873K if the duration of the interaction 12min., is represented on Fig. 2.

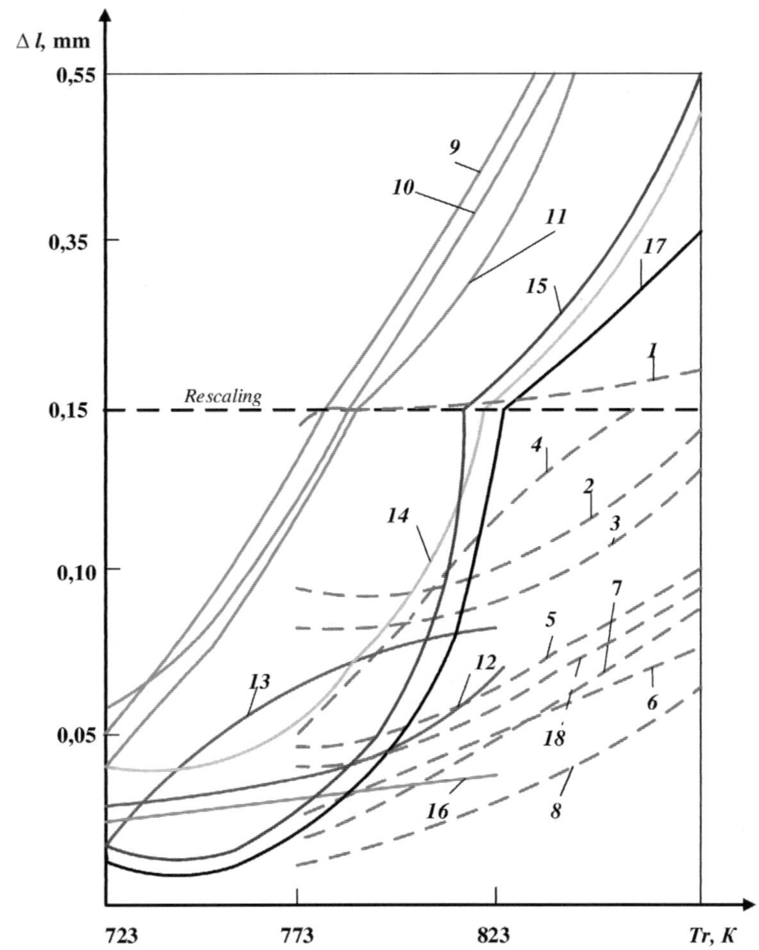

Fig. 2 – The effect of testing temperature on the solubility of tool materials in Zamak melt – 4 – 1: 1 – 3H2V8F; 2 – 4H3VMF; 3 – 5H4SV4MF; 4 – DI 23; 5 – DI 22; 6 – 4H5MFS; 7 – DI 22 ESR; 8 – DI 22EBR; 9 – VK30; 10 – VK25; 11 – VK20; 12 – VK10S; 13 – VK20S; 14 – VK20K; 15 – VK25M; 16 – VK10; 17 – VK25H; 18 – DI 22L.

At lower temperatures (723 – 773K) fine-grained ceramic materials VK10S, VK20S, (curves 12, 13, Fig. 2), and also alloyed with molybdenum, chromium and VK25M, VK25H (curves 15, 17, Fig. 2) in comparison with the steels have almost the same resistance, and with increasing testing temperature in Zamak melt – 4 – 1 they are worse than steel. In this case a clear trend to decrease resistance of WC – Co composites at cobalt increasing in the bundle (curves 9, 10, 11, 16, Fig. 2). So at lower testing temperatures (*Tr = 773K*) alloys VK20S and VK20K have almost the same resistance superior to that of the 1.5 times for alloy VK20. At higher testing temperature (*Tr = 823K*) solubility of alloy VK20K increases dramatically in comparison with VK20S (curves 13, 14, Fig. 2).

Alloying VK25H with chromium increases the resistance of solubility in the melt Zamak – 4 – 1 in the whole test interval temperature (curves 10, 17, Fig. 2). Effect of alloying with molybdenum at elevated temperatures of the melt leads to a reduced resistance of alloy VK25M (curve 15, Fig. 2).

Thus, in terms of solubility in the melt Zamak – 4 – 1 the promising for of manufacturing of details of metal-wire foundry cars CLT and IDRA types are steels 4H5MFS and DI – 22. Considerable reserve for increasing the life of metalwire is the use of cast steels, as well as options of ESR and EBR.

IOP Conf. Series: Materials Science and Engineering **134** (2016) 012033 doi:10.1088/1757-899X/134/1/012033

Metal-ceramic alloy doped with chromium VK25H may be recommended for reinforcement of heavily loaded parts of the press-nodes of hot chamber machines under pressure (press – pistons, cylinder liners).

References

[1] Vitkin A I, Kokorin G A, Grishko A G and Tyukanov P A *1973 Aluminum influence in the zinc melt on the formation of the diffusion layer during hot dip galvanizing of steel* Metallography and heat metal treatment [Vliyanie alyuminiya v tsinkovom rasplave na obrazovanie diffuzionnogo sloya pri goryachem tsinkovanii stali Metallovedenie i termicheskaya obrabotka metallov] chapter 4 pp 60 – 62

[2] Hansen Hi and Anderko K 1962 *The structure of binary alloys* [Struktura dvoynyih splavov] M: Metallurgizdat

[3] Kolesnikov M S, Muhametzyanova G F, Astaschenko V I and Sunagatova E M 2014 *Mechanism and kinetics of wear in metal-wire hot chamber machines of molding under pressure* [Mehanizm i kinetika iznosa metalloprovodov goryachekamernyih mashin litya pod davleniem] Moulder of Russia chapter 6 pp 29 – 32

[4] Kolesnikov M S, Muhametzyanova G F and Zubkov *2015 Optimization of heat treatment modes of steel 4Kh5MFS for metal conduits of hot-chamber die-casting machines according to results of endurance tests in molten* TsAM – 4 – 1 Metallovedenie i termicheskaya obrabotka metallov chapter 11 pp 53 – 56

[5] Muhametzyanova G F and Kolesnikov M S 2015 The *materials and production technologies diecast pressing tool* LAMBERT Academic Publishing p 264

Creating a rigid model of solid end radius cutter in NX

E A Ryabov and S Yu Yurasov

Kazan Federal University, Naberezhnye Chelny Institute
Pr. Mira 68/19, Naberezhnye Chelny, Republic of Tatarstan, Russia, 423812

e-mail: Evgeniy.ryabov@inbox.ru, docfile@yandex.ru.

Abstract. The article describes the sequence of actions in a module NX-CAD to get solid model of the instrument in accordance with GOST 16231-81 "End radius cutter with a tapered shank for processing of light alloys"

Today there are many companies producing tools such as Sandvik, Iscar, DIXI, YUMT, Hofman and etc. Each of these companies produces the directory in which under any specific conditions on the basis of the classifier, you can pick up a tool and recommend treatment regimens, this is true also for the end radius cutter. However, part of the cutting tool at domestic enterprises is often made according to National Standard, as they are given all the design parameters, but they do not give any recommended operating conditions. In view of the relatively high cost of such a tool, the error in the selected cutting conditions can lead to high cost of the final product. To avoid such errors, it is possible to simulate load on the finite element model of a cutter [1] and on the analysis of the data to make recommendations on the use of various cutting conditions.

Solid model of cutter is necessary for the finite element analysis, resulting in CAD systems [2]. Solid modeling in the system Siemens NX is performed in the corresponding module. We create a model of the cutter with diameter d = 12 mm in accordance to National Standard 16231-81 "End radius cutter with a tapered shank for processing of light alloys" [3].

First we construct a sketch in the plane XY (Figure 1), all sizes of face sections according to National Standard [4]:

1) The circumference of the core d=4.8
2) The line of vertical axis and we made it auxiliary
3) circumference d = 4, which applies geometric constraints "point on the curve," is connected to the tip of the tooth, and the "Ref" with a circumference of the core.

4) The line which forms a clearance angle of 20 degrees and a length in the direction X 1.2mm.
5) The circle d = 12, which applies geometric constraints "point on the curve," is connected to the end point of the tooth neck, and regarding the core of the circle
6) Create an array relative to the center
7) fast cutting takes away extra line

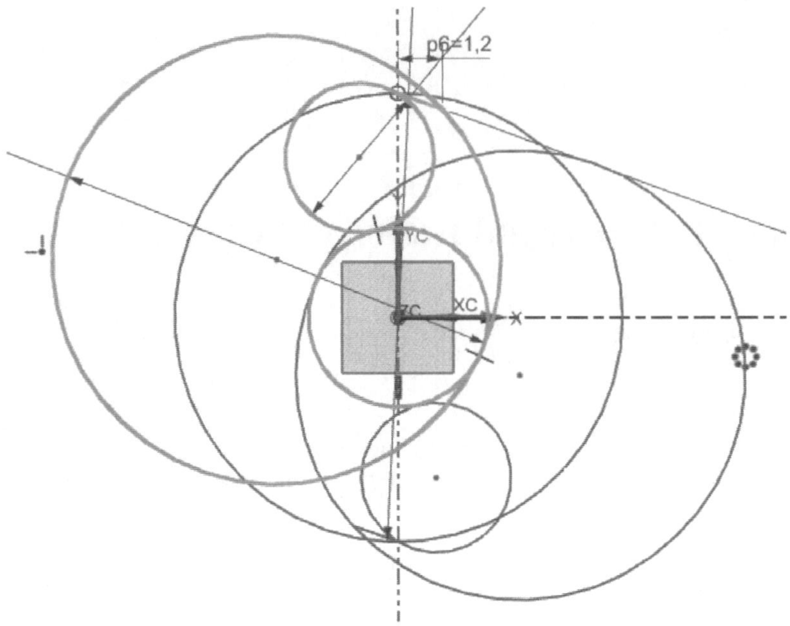

Figure 1. Sketch of the cutter in the face section to trim auxiliary lines.

We create guides on the ZY plane: one as the original tool surface for future spiral, the second line as a core to impart thickening GOST, as well as the centerline of the tool. We define a "Spiral" along the guide axis with the specified size parameter "On the curve of the law," where the curve of the law is the image of the original surface of the tool, the baseline is the tool axis with a pitch P = 65.3 mm. This spiral is the future cutting edge [5]. (Fig. 2).

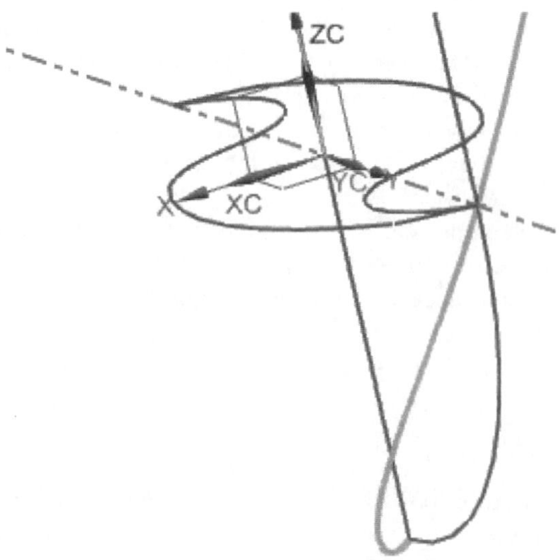

Figure 2. The guide spiral of cutting edge.

To simulate the output of the tool it is necessary to provide radius output in guide for the helix

With the command "sweeping" we build the body of the milling cutter. As the section we select the thumbnail of the end section. We choose two sets of rails, in the first it will be a spiral, in the second it will be an axis section, that is the way section will move along the spiral. However, the cross section will tend to go to the normal to vectors of the guide, in order to avoid this we will define the axis of the reference curve, it will provide a condition under which the section plane at each point of the trajectory will be normal to the axis.

At the end of the cutting part circumference sketch d = 12 mm is created. With boolean operations we will find the intersection of the final approximated profile of the cutting part. With further indentation sketch of a circle with offset and slope we will create the neck of cutters and Morse taper. At the end of the Morse taper we create a sketch rectangle and squeeze the foot. We add "rounding" at the transition elements[6].

The resulting model (Fig. 3) is nearly the same geometry of the actual milling [7] except for the exit site of the second order instrument, but this place is rarely a stress concentrator and almost does not give an error by further engineering calculations.

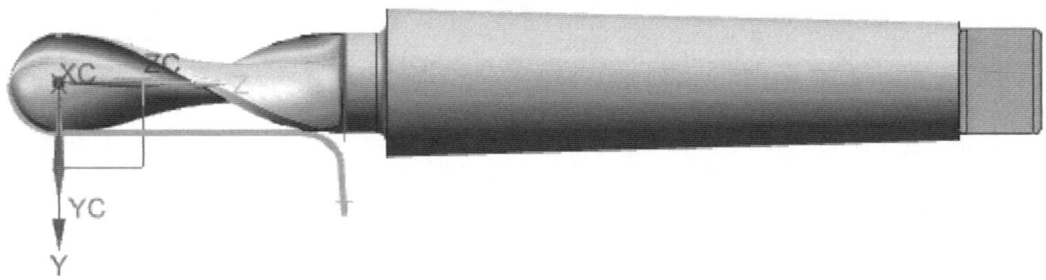

Figure 3. The finished solid model with a dedicated cutter curve-guide spiral. This model is applicable to further finite element analysis.

Bibliography:

[1] L A Simonova, E I Egorova 2015 Development of structural element precedent of technological process in computer-aided design *IOP Conference Series: Materials Science and Engineering* **86** 012017

[2] R M Khisamutdinov, M R Khisamutdinov 2014 Automation system goals for the creation and operation of the tool *IOP Conference Series: Materials Science and Engineering* **69** 012021

[3] 1981 GOST 16231-81. *End radius cutter with taper shank for processing of light alloys* (Enter. 1983-01-01. M .: Order "Badge of Honor" Publishing Standards)

[4] 2006 GOST 25557-2006. *Cones tool. Basic dimensions* (Enter. 2006-06-24. - M .: Federal State Unitary Enterprise "STANDARTINFORM")

[5] Danilov Yu and Artamonov I 2011 *The practical use of NX* (M .: DMK Press) 32p

[6] R M Khusainov, S A Belov, O V Chukhontseva 2014 Diagnosis of cnc machine tools in terms of circular interpolation′s accuracy figure *IOP Conference Series: Materials Science and Engineering* **69** 012022

[7] Golovanov N N 2002 *Geometric Modeling*. (M .: Publisher Physical and mathematical literature) 7p

IOP Conf. Series: Materials Science and Engineering **134** (2016) 012035 doi:10.1088/1757-899X/134/1/012035

Waste of cleaning emulsion sewage as inhibitors of steel corrosion

D D Fazullin[1], G V Mavrin[1] and I G Shaikhiev[2]

[1] Kazan Federal University, Naberezhnye Chelny Institute, Pr. Mira 68/19, Naberezhnye Chelny, Republic of Tatarstan, Russia, 423812
[2] National Research Technological University, Kazan. 420015, 68, Karl Marx street, , the Republic of Tatarstan, Russia

E-mail: denr3@yandex.ru

Abstract. The article describes the corrosion test of steel of the brand 20 in the stratal water. To increase corrosion resistance as a corrosion inhibitor the concentrate waste emulsion of the mark "Incam- 1" was provided. The article presents studies of the corrosion rate with different dosages of corrosion inhibitor in the stratal water. Based on these research results are revealed that the degree of protection of steel is 27% at a dosage of 3.8 g / dm^3.

The problem of corrosion protection of the equipment and facilities of the oil industry has been of paramount importance and highly requires practical solution now.

One of the most common ways to reduce the corrosive effect is the use of corrosion inhibitors. In practice, agents were started to use water-solublemore often, so they are able to create a thin film on the inner surface of the pipeline, to eliminate excessive oiling, do not affect the chemical composition of the oil.

Corrosion inhibitors are divided into inorganic (5% of total amount) and organic (95% of total). Organic corrosion inhibitors are surface-active agents, which are divided into natural and artificial.

Nowadays these corrosion inhibitors as "Amfikor", "SNPCH-1004." are used. These reagents are considered positive, but the disadvantage is the high cost of the reagents. The new chemical agenrdts need to be developed that are not inferior to modern protective properties of corrosion inhibitors and with a low cost.

Exhausted coolants are 3-10% solutions of soluble, which include industrial oils, asidol, ethylene glycol, sodium nitrite, and other substances [1-3]. Both coolants with a short service life and sufficiently resistant to the use within a few months are used in the machine-building enterprises. Coolant of the brand "Incam-1" is the most popular in the Kama region of the Republic Tatarstan. In this case, the emulsion of the brand "Incam 1" as the research object is selected.

IOP Conf. Series: Materials Science and Engineering **134** (2016) 012035 doi:10.1088/1757-899X/134/1/012035

The emulsion of the brand "Incam-1" is a system comprising a mineral oil, emulsifiers, corrosion inhibitors, bactericides and water. The test of cleaning of exhausted emulsion of the brand "Incam-1" is carried out by technological scheme based on membrane method at work [4-6].
Emulsion concentrate obtained after the membrane separation is proposed to use as a corrosion inhibitor.

The speed and extent of the corrosaion protection of metals were determined in accordance to National Standard 9.506-97 by gravimetric method. Metal plates rectangular size 70x35x0.5 mm of steel "Steel 20"as the metal samples were used, from which pipes for the oil industry are made. Simulated formation water for test was prepared by composition that is described in [7]: NaCl - 48 g/dm^3; Ca – 7.4 g/dm^3; Na_2CO_3 – 2.65 g / dm^3; Na_2SO_4 - 0.5 g / dm^3; Na_2S – 0.8 g / dm^3 (corrosive environment №1). Model of saline formation water was prepared in accordance with National Standard 9.506-97 by the following composition: $CaCl_2 \cdot 6H_2O$ - 34 g / dm^3; $MgCl_2 \cdot 6H_2O$ - 17 g / dm^3; NaCl - 163 g / dm^3; $CaSO_4 \cdot 2H_2O$ – 0.14 g / dm^3 (corrosive environment №2). Concentrate of waste emulsion "Incam-1"as corrosion inhibitors was added.

Table 1. Results of tests in corrosive conditions №1

Corrosive environment	Dosage corrosion inhibitors	The average corrosion rate mm/year	Degree of protection, %
Produced water without additives	-	0.628	-
+ 0.2 sm^3 conc. Incam-1	1.5 g/dm^3	0.515	22.0
+ 0.5 sm^3 conc. Incam-1	3.8 g/dm^3	0.493	27.3
+ 1 sm^3 conc. Incam-1	7.7 g/dm^3	0.501	25,3
+ 5 sm^3 conc. Incam-1	38 g/dm^3	0.559	12.4

The results of research have shown that by using the exhausted emulsion concentrate "Incam 1" as corrosion inhibitors (IC) corrosion rate decreases. The optimum concentration of corrosion inhibitors (IC) at which the maximum degree of protection at 27.3%, is determined by the concentration of 3.8 g/dm^3.

Table 2. Results of tests in corrosive conditions №2

Corrosive environment	Dosage corrosion inhibitors	The average corrosion rate mm/year	Degree of protection, %
Produced water without additives	-	0.1210	-
+ 0.2 sm^3 conc. Incam-1	1.5 g/dm^3	0.0974	24.2
+ 0.5 sm^3 conc. Incam-1	3.8 g/dm^3	0.0933	29.7
+ 1 sm^3 conc. Incam-1	7.7 g/dm^3	0.0940	28.7
+ 5 sm^3 conc. Incam-1	38 g/dm^3	0.1014	19.3

According to the results of tests in a corrosive environment №2 the optimal concentration of IC is in the range 3.8-7.7 g / dm^3 of concentration of the waste emulsion "Incam-1". The average corrosion rate in a simulated formation water №1 is 5 times higher than in the water reservoir №2.

Due to the positive test results the waste emulsion concentrate "Incam-1" can be used as corrosion inhibitors (IC) obtained after membrane separation.

Bibliography:

[1] Kostyuk W I 1990 *The cleaning of sewage in engineering enterprises: for engineers and students of technical colleges* (Kiev: Technique) p 118.

[2] Yevdokimov A Yu 1997 *Ecological problems of rational use of used lubricants: abstract diss ... Doctor. tehn.* Sciences. (Moscow) pp 48-50.

[3] Kulakov A T, Gafiyatullin A A, Barylnikova E P, Kulakov A T 2014 *Providing normal conditions of lubricating of diesel engine during its operation* IOP Conference Series: Materials Science and Engineering Volume 69 Issue 1 pp 1-9 Article number 012027 DOI: 10.1088/1757-899X/69/1/012027

[4] Fazullin D D and Mavrin G B 2014 *Cleaning technology of water-based sewage with additional purification by membrane and sorption method* Tehnologies oil and gas №4 (93) pp 3-7

[5] Fazullin D D, Mavrin G V, Melkonyan R G 2014 *Modified cationic membranes for water purification, and their selective permeability* IOP Conference Series: Materials Science and Engineering Volume 69 Issue 1 pp 1-4 Article number 012032 DOI: 10.1088/1757-899X/69/1/012032

[6] Fazullin D D, Mavrin G V, Sokolov M P 2015 *Chemistry and Technology of fuels and Oils* Vol. 51 № 1 pp 93-98 DOI: 10.1007/s10553-015-0579-8

[7] Vinogradov S N, Volchikhin W I, Shirina Ye W, Meshcherskaya A S *2008 Research of the corrosion resistance of structural steel in produced water Engineering* Machine building and machine science №4 pp 139-144

IOP Conf. Series: Materials Science and Engineering **134** (2016) 012036 doi:10.1088/1757-899X/134/1/012036

Algorithms development of making special techniques in APQP manufacturing process of automotive components

D T Safarov, K A Fedorov and A I Ilyasova

Kazan Federal University, Naberezhnye Chelny Institute
Pr. Mira 68/19, Naberezhnye Chelny, Republic of Tatarstan, Russia, 423812

Abstract. Special techniques algorithms are worked out, such as MSA, preparation of management plans, SPC in the preparation of manufacturing automotive components that improve its effectiveness.

Features of automotive industry are large volumes of products. The vehicle is a structurally complex object. Head factory itself is not able to produce many automotive components and is forced to work with no less than 500 suppliers. And if one of them fails to supply (not provide the quality or hamper deadlines), problems can arise in the entire corporate system. The complexity of working with suppliers is determined not only by their number, but also by a variety of automotive components. These are both large aggregates and all electronic components and a variety of materials, standard products. Therefore, the headquarter plant is extremely important to work with any supplier by one and the same technique. Now this is a production part approval process (PPAP) [1].

American motor corporations developed it in the early 1990-s, and today it is standardized on the global scale. The supplier must submit a package of 19 documents and samples (Fig. 1) to the consumer after completing pre-production of automotive components previously not supplied, as well as in the event of provider significant changes in the production. The data set should confirm to the consumer the quality of the components, the stability of its production processes, as well as the effectiveness of the quality assurance system supplier. A standardized version of the procedure is fairly common. All the world's leading corporations develop on its own corporate system based on the approval of the production, that is, require suppliers to provide more detailed information.

Unfortunately, Russian motor corporations continue to operate facing mainly to the domestic market. From the time of administrative economy it inherited tradition to accelerate the start of series production without paying attention to the quality of its preparation.

The basis for production management in our enterprises is still the Taylor system, which was enough to assign standardized requirements and monitor their implementation. And long-running managers and specialists, and current students, who are still trained with the same Soviet educational standards that are designed to train future plans for batch production performers.

IOP Conf. Series: Materials Science and Engineering **134** (2016) 012036 doi:10.1088/1757-899X/134/1/012036

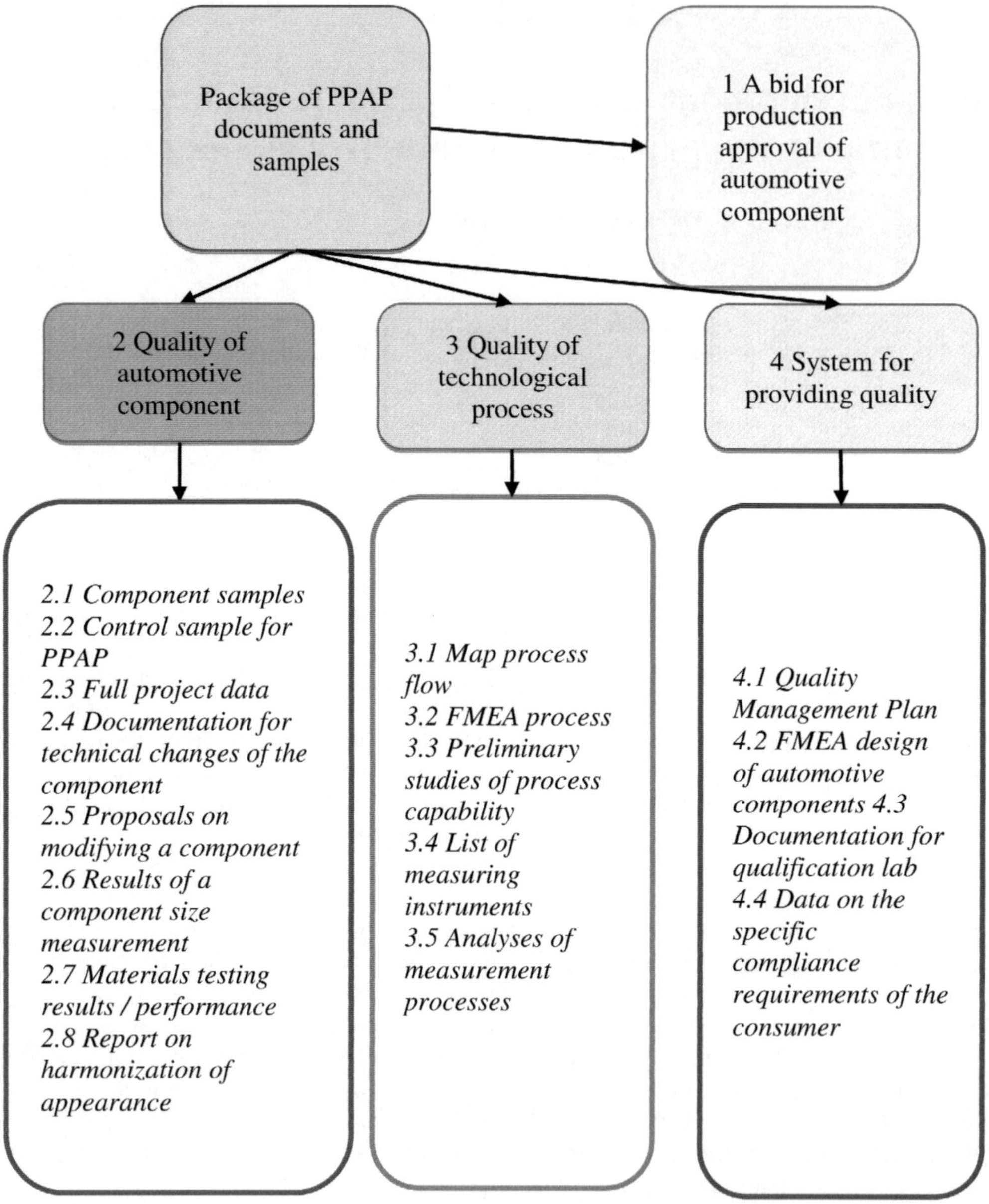

Fig. 1 Set of documents for approval of the production of automotive components.

For our corporation this procedure has not yet been worked out. The owners of the main plant are not sure of the prospects for development. They are important only for today's high profitability. Main factories do not pay enough attention to the objectivity of the data set. Russian specifics of preparation kit are that all suppliers do not concern about the objectivity of the data set. Instead of cooperation to strengthen the position of consumers in a competitive market environment they solve a simpler problem: to stay at any cost among the existing suppliers. And for this is considered valid to cover existing problems in the production by the formal report, not bothering tracking information from the technical documentation in the form of PPAP.

If the pre-production made with omissions, (it is studied in most universities only as a branch of the discipline "Production and Management"), then the process documentation and maintenance of PPAP documents in fact cannot reflect the real state of production.

For a small enterprise specialized in manufacturing parts when PPAP situation occurs, it is suggested for the representative consumer (he provides interaction) to prepare a work plan agreed with the customer.

On the plan basis order form is prepared for each performer. In the context of a small business it is likely to be a representative of the consumer itself, technologist, quality management representative. Technical work on the design of the kit can be performed by the assistant director.

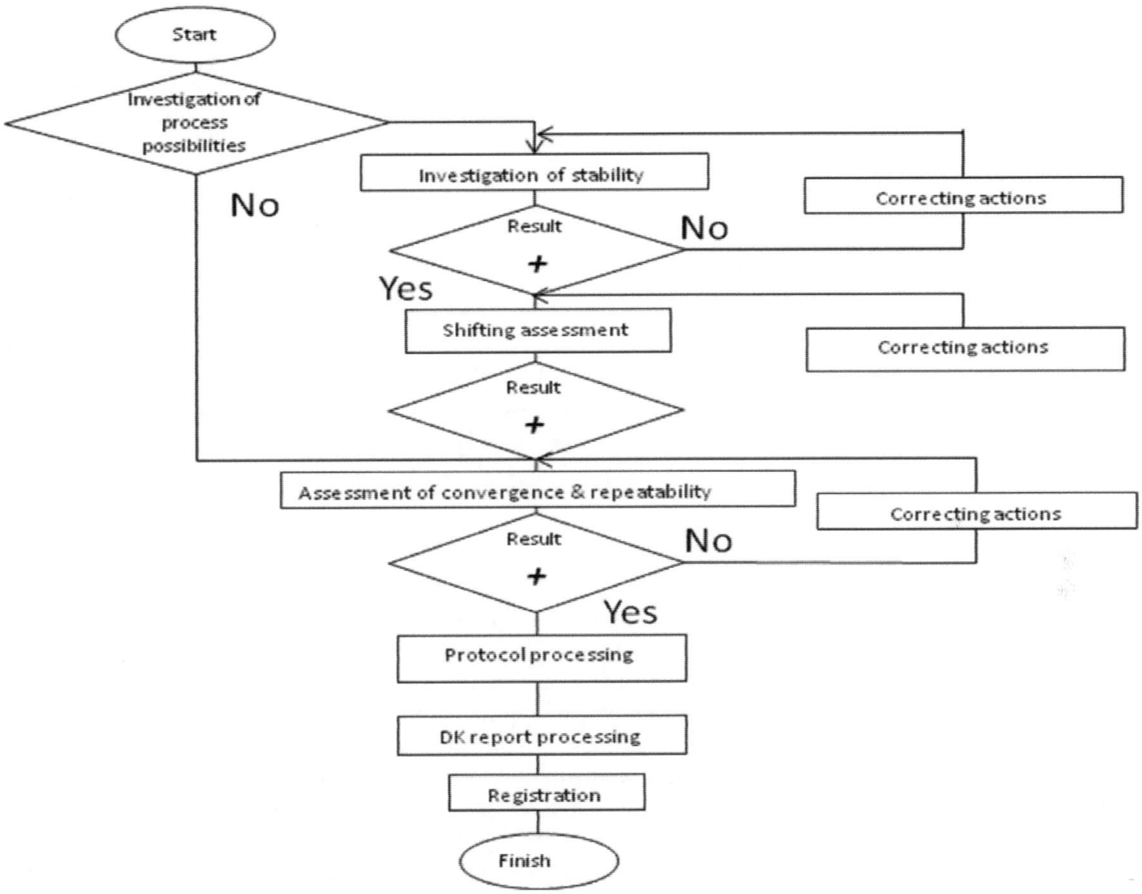

Fig. 2 Algorithm analysis of measurement processes.

IOP Conf. Series: Materials Science and Engineering **134** (2016) 012036 doi:10.1088/1757-899X/134/1/012036

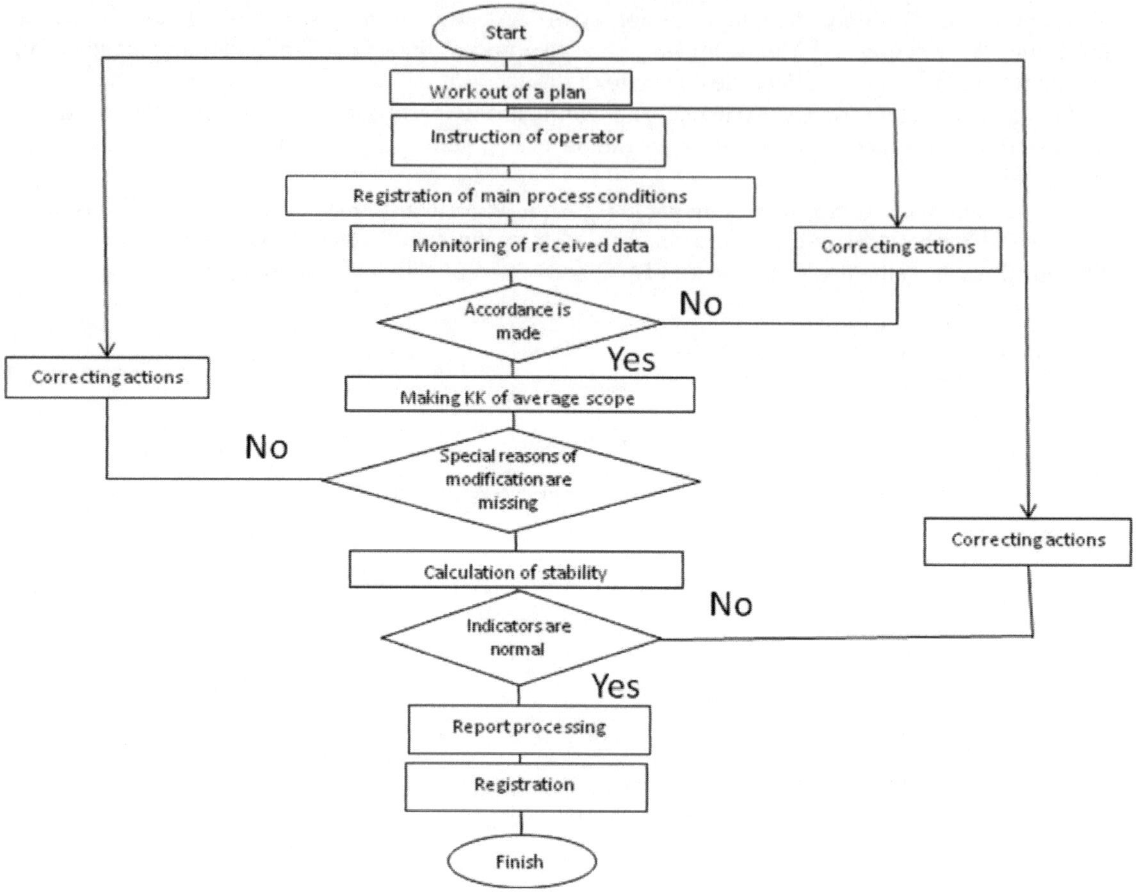

Fig. 3 Algorithm of technological process capabilities analysis.

To increase understanding of information value by professionals to manage quality it is proposed to consider a set of PPAP in relation to the reconciliation of mandatory detention of all documents on management of production processes. The fact that it is quite often to do this, it is important to do this as quickly as possible, avoiding mistakes that could lead to a denial of production approval. During pre-production a number of special procedures is done in accordance with [1, 2, 3, 4]. В том числе особенное на это этапе особое внимание следует уделить проектированию специального инструмента, повышающего производительность [5].

Let's consider algorithms which perform the most complex practical commissioning (Fig. 2.3). So procedures confirm the suitability of the measurement of quality key indicators which are accomplished in the following sequence: stability is investigated and then displacement of the measurement process is estimated and only in case of success of both procedures of convergence and repeatability is analyzed and then a protocol analysis and registration is issued.

Another procedure is a confirmation of statistical quality (stability) of technological process (SPC). It requires initial alignment of the content of routings, management planning, briefing operator registration process conditions - the failure of which may result in the impossibility of the determination of corrective actions in case of discrepancies.

Наибольшую результативность в улучшении качества продукции можно достичь применяя одновременно с SPC - методику диагностирования по показателям технологической точности, когда степень идентификации обрабатываемой детали значительно повышается [6, 7]. С применением методик непосредственного диагностирования состояния узлов оборудования [8].

IOP Conf. Series: Materials Science and Engineering **134** (2016) 012036 doi:10.1088/1757-899X/134/1/012036

Algorithms and other mandatory procedures are developed, such as the development of process flow charts, plans of quality management, action on a result of the production approval by the consumer.

The entire set of algorithms are the basis of existing management processes in enterprises, which reduce errors and improve the effectiveness of the procedures for the preparation of production.

References:

[1]State standard R 51814.5-2005. Quality management system in automotive industry. Analysis of measuring and control processes. (Moscow, Publishng of standards) 63p.

[2] State standard R 51814.3-2001 Quality systems in the automotive industry. Methods for statistical process control. (Moscow, Izdatelstvo standartov) p.35.

[3] State standard R 51814.4-2004 Quality management systems in the automotive industry. Production approval of automotive components. (Moscow, Izdatelstvo standartov) p.43.

[4] State standard R 51814.6-2005 Quality management systems in the automotive industry. Quality management in planning, development and manufacture of automotive components. (Moscow, Izdatelstvo standartov) p.42.

[5] Kondrashov A G, Safarov DT, Davletshina G K 2015, *Bit for the core drilling of nonmetals (Article) Russian Engineering Research*, Volume 35, Issue 8, 26 August Pages 617-618

[6] Kas'yanov S V, Safarov D T. 2004 *Diagnosis of technical state of equipment and tools according to indices of technological accuracy (Article)* Avtomobil'naya Promyshlennost Issue 5, Pages 24-28.

[7] Balabanov I P, Balabanova O N, Groshev A V, *Formation of initial data of the workpiece batch in simulation modeling precision forming*, IOP Conference Series: Materials Science and Engineering, Volume 86, Issue 1, 26 June 2015,

[8] Khusainov R M, Belov S A, Chukhontseva O V 2014, *Diagnosis of cnc machine tools in terms of circular interpolation's accuracy figure*/R M Khusainov, S A Belov, IOP Conference Series: Materials Science and Engineering, Volume 69, Issue 1

Displaying structural property and inheritance of cast iron surfacing on steel base

I.A. Shveev

Kazan Federal University, Naberezhnye Chelny Institute
Pr. Mira 68/19, Naberezhnye Chelny, Republic of Tatarstan, Russia, 423812,

e-mail: shveev_andrey222@mail.ru

Abstract. Graphite inclusions heredity in deposited layer from remelted special cast iron billets was established. The possibility of controlling the structural state and the quality of the deposited layer due to technological parameters of welding and heat treatment of parts is shown. Ways of improving cast iron wear resistance durability are proposed.

A fundamental factor in the stabilization and achieving a high level of parts resource is to improve the quality of the surface layer, as it is in all cases the most loaded. Depending on the type and degree of loading and the nature of the medium, etc. for most efficient technologies are recognized hardening contact surfaces based on the coating, doping, thermal effects and deformation, as well as by their combination. In some cases, an undoubted advantage is the use of alloys with high specific properties. The key parts of the engine, determining its durability are part timing, which are manufactured with a wide range of materials.

Reliability and durability to operate a pair of "camshaft-tappet" is directly dependent on the performance of the cam contact surface of the camshaft with the end of the valve tappet. A characteristic feature of this pair is the case of the contact of dissimilar materials, namely steel hardened cemented surface of the camshaft with cast iron tappet. Special demands to the contact surface of the tappet representing special cast iron weld on a steel base parts (Figure 1). These requirements include hardness values (not less than 61 HRC) and structure (ledeburite and sections of martensite with residual austenite). In addition, a small amount of cementite and dotted, evenly distributed graphite is allowed. A disadvantage of the tappets in operation is a dispersion of resistance (uneven wear, chipping, etc.), despite the fact that they are manufactured and hardened under the same conditions of production, having high and a predetermined structural state (Figure 2). However, the stability of individual elements of the structural state of iron on the part such as graphite inclusions are not always achieved and important role here is the macro- and microstructure inherited from cast billet used for surfacing.

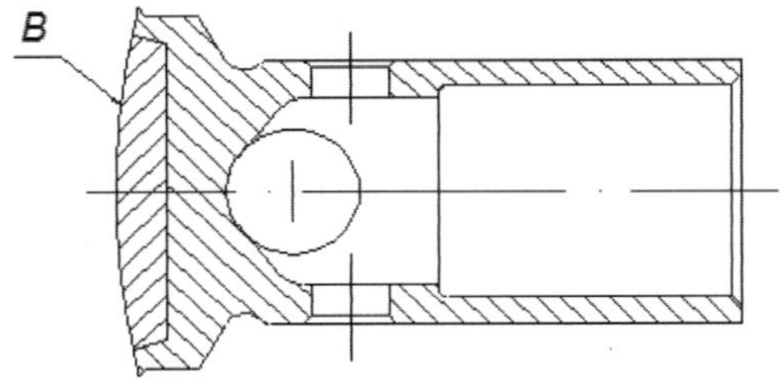

Figure 1 - Design of the valve tappet

The end of the valve tappet (fig.1), working on the cam of the camshaft is surfaced by special cast iron of the following composition (wt.%): C-3.1 ÷ 3.4%; S-2.1 ÷ 2.35%; Mn-0.8 ÷ 1.0%; Ni-0.4 ÷ 0.75%; Mo-0.4 ÷ 0.60%; S≤0 / 1%; R≤0.2%. Cast billet of Ø10 × 48mm is used for surfacing and a weight from 27.0 up to 28.5gramm. Surfacing was carried out on UNT-1 with the heating TVC of the steel parts up to 1100 ° C, and cast iron billets to 1250°S.

Cooling the items is carried by water under pressure from 1.0 to 2.0 atm., fedding into the inner cavity of the part. After the final machining of parts subjected to oxidation in a moist nitrogen atmosphere of the US made shaft furnace 8.12 / 6 at a temperature of 450 ± 5 ° C during 4 hours. Microphotometer MFS-51 and analyzers AN-7529 and AS 7232 have been used for chemical and spectral analysis of alloys.

Figure 2 - The type and nature of the valve tappet surface wear

The methodology of the macro and micro studies based on the use of standardized methods for monitoring and testing using reliable and licensed equipment. Control patterns before and after etching of 4% solution of picric acid was performed on microscope "Neofot-2" and "Epitip" at 100 and 400 *enlargement. The hardness was measured by Rockwell and micro hardness of the individual structural components was measured on "Durimet" (Germany). The content, appearance and character of the distribution and structure of the graphite layer deposited on the parts was evaluated in accordance with GOST 3443-87.Macroanalysis performed after etching of deposited items in 70-85°C for 10 minutes in a 50% aqueous hydrochloric acid. Detection of pores and cracks in the finished products is carried by magnetic and powder technique on a flaw detector DN200P.

Currently used for surfacing of special cast iron has good wear resistance, but the durability of these parts in the operation of the alloy is limited by the presence of undesirable macro- and micro-defects. To those clusters should include graphite inclusions, pores, cracks, presence nonmartensitic conversion products and residual austenite. Despite a wide range of research in the field of structure and graphitization iron undoubted interest are issues relating to heredity structure and properties of the alloy after remelting. We studied the influence of the silicon content as essential graphitizing elements, the structure and condition of the metal substrate graphite nodules in the deposited layer formed with a high crystallization rate and subsequent cooling of the remelted iron. It is found that the variation of the silicon content in the iron from 0.98 to 2.2% retains high hardness of the deposited layer (60-61 HRC), the structure of which is composed of cementite, ledeburite, martensite, residual austenite and graphite. Under reduced silicon content (0.98 and 1.3%) in the alloy is observed grinding martensite-austenite sections and according to X-ray structure analysis of residual austenite is 3-5%. Along with dotted and uniformly distributed graphite in an increased silicon content in the iron (2.2%) is observed further selection eutectic (in the form of rosettes) graphite. Silicon has a marked influence on the properties of structural components in the deposited iron. Increasing its content from 0.98 to 2.2% contributes to the continuous improvement of ledeburite micro hardness which reaches 1080HV. Regarding martensite we revealed that a maximum micro hardness occurs when the content of the deposited layer is 1.95% of silicon. Changes in the content of this element in the direction of increasing or decreasing from this value decrease the micro hardness of martensite.

Key attention is paid to succession layer of weld shape and distribution of graphite from the cast billets. It was found that the melting temperature range of cast billets and shutter speed of the metal in the liquid state has a significant impact on the character of graphite inclusions in the deposited layer of detail. The results of metallographic studies and long production experience convince us that there is a connection between the state of graphite in the workpiece and the weld layer of the part and give grounds to assert that there is a real opportunity to manage this phenomenon by changing the process parameters surfacing. The presence of lamellar graphite and female connector in the workpieces not only reduces their hardness, but also retains the undesired state of graphite in the deposited layer, regardless the temperature and the boiling time of iron in the test range. Having dot rosette with inclusions of graphite in the white iron contributes to the hardness 46-54 HRC blanks and such a state is retained in the deposited graphite layer, if the boiling point of iron is 1230°C and lower. Increasing the exposure time to 12 seconds of the alloy in the liquid state at a temperature 1230°C as an increase of the boiling point of iron up to 1250°C favor for the dissolution of the graphite clusters and uniform distribution of dotted of its discharge into the structure of welding. The results are in good agreement with the opinion of well-known scientists [1, 2, 3], indicating that one of the effective methods of controlling heredity is the melting of iron, followed by a certain overheating.

A noticeable effect on the properties of cast iron has a special cooling process of parts after surfacing. Reducing the pressure of water supplied into the parts for cooling from 1.2 atm. to 0.4 atm. decreases microhardness of martensite to 504-530 HV while maintaining ledeburite microhardness on the level 1000-1100 HV. One of the reserves of increasing martensite microhardness up to 700-725 HV, increase of steel base hardness 47-50 HRC and the deposited layer up to 61-62 HRC, as well as reducing the depth of the pores is the introduction of an additional external cooling under the plate of details after surfacing (table. 1 and table 2). In addition to this effect as a result of the accelerated cooling is observed directional solidification of the alloy throughout the volume of the deposited layer and create a positive environment for the formation of a dotted, stable, evenly distributed graphite surfacing.

Table 1. Properties of the parts and the deposited layer after different cooling conditions.

Way of cooling	Hardness, HRC		Microstructure of deposited layer	
	Steel base	Deposited layer	Base	State of graphite
Water delivery into the part	45-47	59-60	Cementite, ledeburite, martensite areas and	Dotted and eutectic

			austenite	
With additional external cooling	45-50	61-62	Cementite, ledeburite, martensite areas and austenite	Dotted, evenly distributed

Note: Pressure of water supply is 1.2 atm., temperature is 17^0C.

An additional condition for improving wear resistance of parts can serve increasing the antiscoring property of working surface, which is achieved by thermal oxidation. Particular attention deserves oxidation technology of valve tappets, which would significantly increase the wear resistance of parts (Table 2).

The technology provides cleaning and degreasing parts in hot (70-80^{0}C) KM-1 detergent by drying, heating to 300^{0}C in a nitrogen-based protective atmosphere, then the feed of nitrogen is stopped and heated to 450^{0}C in a humidified atmosphere created by the putting water into the furnace in an amount of 1.0-1.4 l / h upon 1 м^3 of the workspace, kept for four hours and then cooled to 180-200^{0}C in this environment and then in air.

Table 2. Effect of oxidation parameters on properties of parts.

Oxidation parameters		Thickness* of oxidation layer, mkm. (steel, cast iron)	Magnitude of wear, ** mkm.
400	1,2	5/4	10,5
	1,4	5/4	10,5
450	0,5	4/2	17,5
	1,0	5/4	11,0
	1,2	5/4	11,5
	1,4	6/5	10,0
	2,0	7/5	19,5 (chalking)
500	1,0	6/4	12,5
	1,2	6/4	13,0
	1,4	7/5	10,0

Note: * - oxide layer is formed at 150 minutes of exposure
** - magnitude of wear is determined after 53 hour test on the stand

Conclusion.

1. Heredity of graphite inclusions is established and shows the ability to manage the structural state of deposited layer by technological parameters of special cast iron surfacing and heat treatment of parts.
2. Workpiece surface purity and the cooling rate of the parts after surfacing have a direct effect on the formation of pores and their depth in the deposited layer.
3. Additional increase in wear resistance of the deposited layer is achieved by thermal oxidation of the parts.

References:

[1] Girshovich N G 1966 *Crystallization and properties of cast iron in castings* (Moscow: Mashinostroyeniye) p 368

[2] Kushnirsky A S 1936 Influence of heredity, source materials graphite forms, submicroscopic inclusions and nature of slag on structure and mechanical properties of iron casting No 2 (Moscow: Litejnoye delo) p 13-18
[3] Akst E R 2015 Structural and phase transformations in cast irons when implanted into the surface layers of nitrogen ions *IOP Conference Series: Materials Science and Engineering* ***86*** *012038*

IOP Conf. Series: Materials Science and Engineering **134** (2016) 012038 doi:10.1088/1757-899X/134/1/012038

Prediction of fracture based on an assessment of "damage" of a metal by plastic deformation

R V Shibakov and V G Shibakov

Kazan Federal University, Naberezhnye Chelny Institute
Pr. Mira 68/19, Naberezhnye Chelny, Republic of Tatarstan, Russia, 423812,

e-mail: vladshib50@gmail.com

Abstract. Using the hypothesis of damage summation and the linear law the accumulation of such a method is proposed for predicting the fracture with large plastic deformation.

Modern views on the mechanism of failure during plastic deformation are based on the experimentally established fact "loosening" of the material associated with the origin, development and merging of damage (pores) with the growth of the strain with the subsequent formation of microtrain, ultimately leading to destruction[1,2].

In the framework of a phenomenological approach, considering the damage without specifying its physical form, in categories in continuum mechanics, we will create a model of the fracture of materials by cold plastic deformation.

The damage will be characterized by a certain scalar quantity Ψ, which displays the degree of exhaustion of ductility with increasing strain. The value of Ψ in the initial (undeformed) state Ψ=0, Λ=0 (Λ is the degree of shear deformation of the considered material particle of a deformable body) $\Psi p=1, \Lambda=\Lambda_p$ at the moment of fracture, when the ductility of the metal material in the particle is exhausted, Ψ<1 in the range of $0<\Lambda< \Lambda_p$ .The function Ψ=Ψ(Λ) displays the pattern of accumulation of Ψ as the exhaustion of ductility due to the growth of the indicator strain Λ. K=K(Λ) – dependence, showing the evolution of stress-strain state in a neighbourhood of the considered material particle (path deformation).

Such a representation of the history of deformation and possibilities of plastic deformable material in a single system of coordinates Λ=Λ(K) allows to estimate the damage of Ψ as a function of the form

$$\Psi = [K(\Lambda), \Lambda_p(K)],$$

and admits a geometric interpretation (Fig. 1) development of fracture process.

IOP Conf. Series: Materials Science and Engineering **134** (2016) 012038 doi:10.1088/1757-899X/134/1/012038

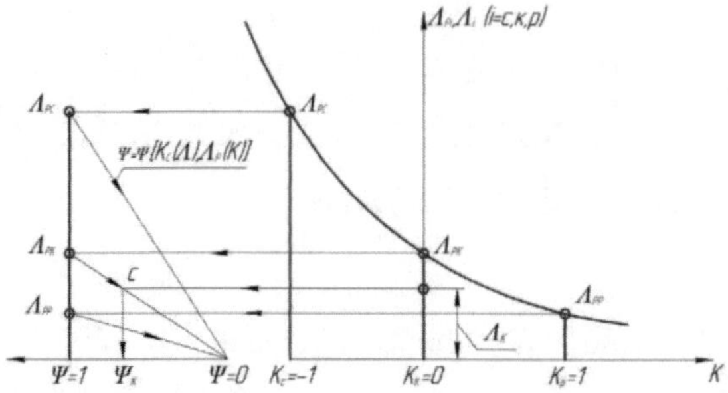

Fig. 1 Diagram to determine the patterns of damage accumulation is $\Psi = \Psi[K_i(\Lambda_i), \Lambda_p(K)]$ when the deformation by compression, torsion, stretching for K=const.

In the right part of Fig. 1 is shown conventionally the diagram of plasticity of the material of $\Lambda_p=\Lambda_p(K)$ in the representation of Bridgman. In the right part of the picture – according to $\Psi = [K(\Lambda), \Lambda_p(K)]$, – reflecting a pattern of damage accumulation for a linear law.

During construction of chart plasticity of $\Lambda_p=\Lambda_p(K)$ are chosen such kinds of tests in which the score, if possible, should not change during the test, i.e., K≈const. In Fig. 1 are schematically presented data on the tensile, compression, torsion, tension, in which the index schema of the stress state are respectively equal to K_c=-1, $K_к$=0, K_p≈1, and the limit degree of deformation at the time of destruction is marked as $\Lambda_{pc}>\Lambda_{pк}>\Lambda_{pp}$. Line $\Lambda_{pc}0, \Lambda_{pк}0, \Lambda_{pp}0$ (right part of Fig.1)characterizes the pattern of change Ψ in compression, torsion, respectively, when taking into account the hypothesis of linear damage accumulation. The definition $\Psi_{\Lambda_к}$ accumulated in the process of deformation distortion, such as twisting to the value of the degree of deformation of $\Lambda_к$ can be carried out graphically, as the intersection of a line $0\Lambda_{pк}$ with a line drawn parallel to the abscissa of the point $\Lambda_{к.}$, From the simirarity $\Delta 01\Lambda_{pк}$ and $\Delta 0\Psi_к C$(left part of Fig.1)results that

$$\Psi_{\Lambda_к} = \Lambda_к / \Lambda_{pк}, \quad (1)$$

the amount of potential to fracture damage, which means the residual resource of plasticity is equal.

$$\Psi_{ок} = 1 - \Psi_{\Lambda_к}, \quad (2)$$

In Fig. 2 a, b presents a calculation scheme of determination of damage and limit the degree of deformation Λ_Σ for the set of monotonic deformation:

a) $\Lambda_к$ ($K_к$=const) + Λ_p (K_p =const), torsion +tension to failure

б) Λ_p (K_p=const) + $\Lambda_к$ ($K_к$ =const),tension +torsion to failure

In the case of a) deformation at the beginning is performed by torsion under a "softer" scheme of the stress state $K_к$ =0=const, the definition $\Psi_{\Lambda к}$ и $\Psi_{0к}$ shown earlier in figure 2 and are calculated with relations (1), (2).

Further, in the present case, the deformation by stretching, is performed with more "rigid" scheme of the stress state of K_p =1=const at a higher rate of damage accumulation is described by the line 0 Λ_{pp} . Allowed to destruction, deformation stretching Λ_p taking into account the relations (1), (2) and the similarity of triangles $0\Lambda_{pp}1$ and ΔΛppΨKC' will be equal to

IOP Conf. Series: Materials Science and Engineering **134** (2016) 012038 doi:10.1088/1757-899X/134/1/012038

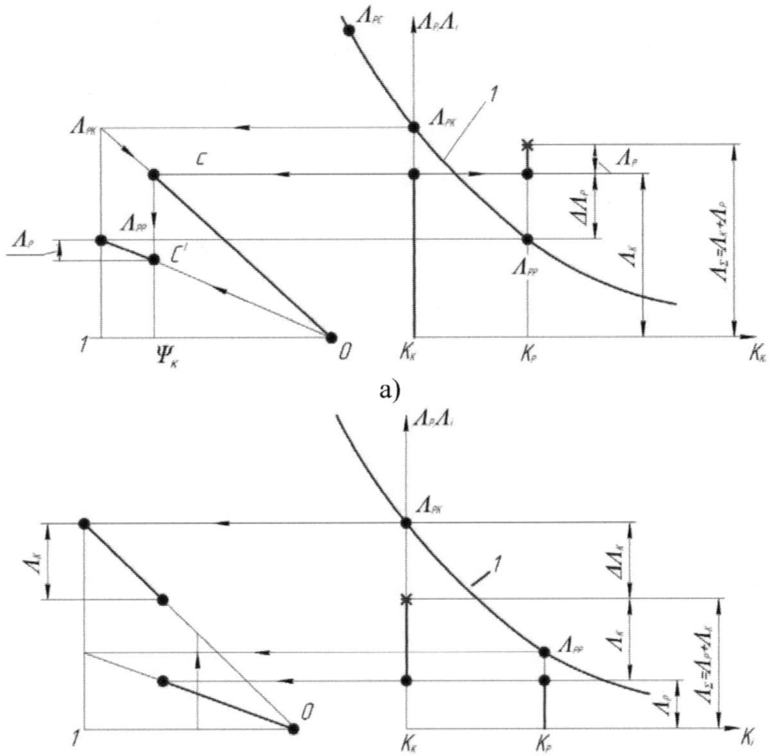

Fig. 2 Calculation scheme for the definition of the parameter "damage" (Ψ) of a steel and limit the degree of deformation (Λ_Σ) depending on the way of deformation K=K(Λ);a) the path of deformation – torsion to$\Lambda_{K=0=const}$+ elongation till fracture;Λ_{Σ}=Λ_K + Λ_P>Λ_{PP};b) the way of deformation – stretching to$\Lambda_{PK=1=const}$ + torsion till deformation $\Lambda_{K=0=const}$; Λ_{Σ}=Λ_P + Λ_K<Λ_{PK},1- chart of plasticity

$$\Lambda_p = (1 - \frac{\Lambda_\kappa}{\Lambda_{p\kappa}})\Lambda_{pp}, \tag{3}$$

Before the final destruction of the degree of deformation $\Lambda_{\Sigma_{(\kappa+p)}}$ for path deformation for case a) is equal to

$$\Lambda_{\Sigma_{(\kappa+p)}} = \Lambda_\kappa + \Lambda_p > \Lambda_{pp}, \tag{4}$$

Thus, despite the fact that the completion of deformation in this case is stretching, where the ultimate degree of deformation of equals the Λ_{pp} the total degree of deformation Λ_Σ the destruction will be more Λ_{pp}на $\Delta\Lambda_p$ as pre-deformation was carried out with a "softer" scheme of the stress state.

The destruction in Fig. 2 is marked with *, which is located above the curve plasticity of Λ_p=Λ_p(K).

Similarly, Fig. 2 b) was determined by accumulated damage ψ_P, the total plastic deformation up to fracture $\Lambda_{\Sigma(p+\kappa)}$and Λ_K in a different way of deformation (scheme b) (at the beginning of tension on the Λ_P and subsequent deformation by torsion to failure). From the obtained results it follows the following pattern: when the paths of deformation K=K(Λ) emerging from a "soft" K_H schemes of the stress state in the direction of "hard" schemes accumulated by the time of the destruction of the limiting degree of deformation Λ_Σ is greater than the limiting degree of deformation on the plasticity

IOP Conf. Series: Materials Science and Engineering **134** (2016) 012038 doi:10.1088/1757-899X/134/1/012038

chart in K_K corresponding to the moment of destruction. The reverse pattern is observed when deformation paths developing from the "hard" K_H schemes of the stress state in the direction of "soft" one.

$$\begin{array}{l|l} \Lambda_\Sigma > \Lambda_{PK_K} & \text{where } K_H < K_K \\ \Lambda_\Sigma < \Lambda_{PK_K} & where\ K_H > K_K' \end{array} \quad (5)$$

where K_H and K_K are the indicators of the scheme of stress state for the beginning and end of deformation.

Experimental evaluation of limit deformations of destruction on the routes of deformation showed the similarity in the range of 10-15% from the calculated values.

References:

[1] Migachev B A 1994 *Features of damage accumulation during the hot deformation of metals. –* Metals, No. 4, pp. 135-140.

[2] V G Shibakov, D L Pankratov, R V Shibakov 2014 Connection between machines durability during operation and the complex of physic-mechanical properties, formed during the production of critical parts *IOP Conference Series: Materials Science and Engineering* **69** 012040

IOP Conf. Series: Materials Science and Engineering **134** (2016) 012039 doi:10.1088/1757-899X/134/1/012039

A mathematical model of the controlled axial flow divider for mobile machines

V L Mulyukin, D L Karelin and A M Belousov

Kazan Federal University, Naberezhnye Chelny Institute
Pr. Mira 68/19, Naberezhnye Chelny, Republic of Tatarstan, Russia, 423812,

karelindl@mail.ru

Abstract. The authors give a mathematical model of the axial adjustable flow divider allowing one to define the parameters of the feed pump and the hydraulic motor-wheels in the multi-circuit hydrostatic transmission of mobile machines, as well as for example built features that allows to clearly evaluate the mutual influence of the values of pressure and flow on all input and output circuits of the system.

With the stricter requirements on environmental and rising fuel prices, the problem of improving the energy efficiency of mobile machines (MM) is becoming more relevant.

Today, one of the ways to improve efficiency is a combination of a power unit with an electrodynamic, or a hydrostatic transmission. For high MM is more appropriate to use (in terms of efficiency) a hydrostatic type of transmission (based on the volume of the hydraulic drive) beyond electrodynamics in specific capacities in 2,5-3 times.

In the composition of the hydrostatic transmission are traditionally used adjustable hydraulic pump with one or more hydraulic motors, as well as guiding and regulating equipment. There are possible circuits connecting the hydraulic motors to the wheels via mechanical transmission adder, and circuits connecting directly to the wheels in which each wheel can be controlled independently by means of valves with proportional electromagnetic control [1,2,3,4,5]. It is also possible with the use of an adjustable axial flow divider, operating on the principle of axial piston hydraulic machines, which in conjunction with control system, allows significantly improve the mobility and the controllability of the vehicle, and significantly improve the efficiency of the entire power plant as a whole.

The use of unregulated volume flow divider in the cooling system of MM is considered in work [6]. This article is devoted to modeling of processes in proposed a regulated axial flow divider.

For calculation of the required pump pressure, for a known pressure drop in each line, which is connected with the working section (the direct problem), and calculation of the magnitude of the pressure in each working section at a known pressure of the pump (the inverse problem), we use the balance of the hydraulic input power and outputs of the controlled axial flow divider (see Fig. 1) [6]:

IOP Conf. Series: Materials Science and Engineering **134** (2016) 012039 doi:10.1088/1757-899X/134/1/012039

$$Q_n \cdot P_n = \sum_{i=1}^{k} (Q_i \cdot \Delta P_i), \tag{1}$$

where Q_n– pump,м^3/c; P_n– the discharge pressure of the pump, Pa; Q_i– the flow passing through the i-th section,м^3/c; ΔP_i – the magnitude of the pressure drop in Magistrali-th section, Pa; k – number of working sections of the divider.

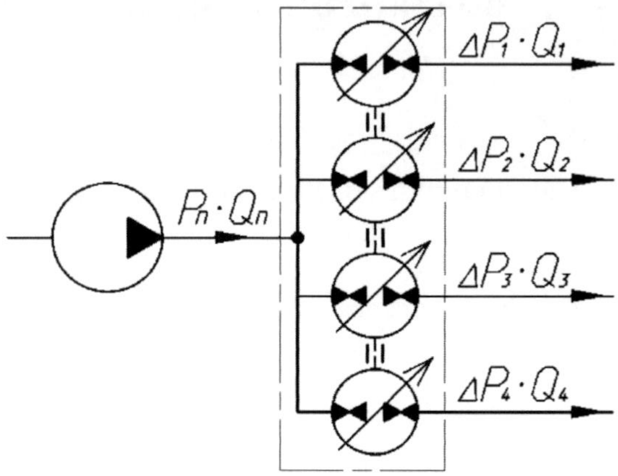

Figure 1 – Scheme for the compilation of balance of the hydraulic power.

Given the fact that the average value of the flow through an axial-piston hydraulic machine according to [7], is defined as $Q_{cp} = D \cdot tg(\gamma) \cdot \left(\pi/4 \cdot d^2\right) \cdot n \cdot z$, and the pump flow is fully distributed to sections of the flow divider, the balance of power, for example, for a 4-section flow divider can be written in the following form:

$$D \cdot \left(\pi/4 \cdot d^2\right) \cdot n \cdot z \cdot \left[tg(\gamma_1) + tg(\gamma_2) + tg(\gamma_3) + tg(\gamma_4)\right] \cdot P_n =$$
$$= D \cdot \left(\pi/4 \cdot d^2\right) \cdot n \cdot z \cdot \left(\Delta P_1 \cdot tg(\gamma_1) + \Delta P_2 \cdot tg(\gamma_2) + \Delta P_3 \cdot tg(\gamma_3) + \Delta P_4 \cdot tg(\gamma_4)\right) \tag{2}$$

where D– the diameter of the location of the axes of the plungers in the cylinder block, м;d – the diameter of the plunger, м; n – the shaft rotation frequency divider, Rev/s; z – the number of plungers in the cylinder block; $\gamma_1 ... \gamma_4$ – the angles of inclination of the washer sections (see Fig. 1).

After mathematical transformations we get finally:

$$P_n = \sum_{i=1}^{k} \left(\Delta P_i \cdot tg(\gamma_i)\right) \Big/ \sum_{i=1}^{k} \left(tg(\gamma_i)\right). \tag{3}$$

Equation (3) allows us to solve the "direct" problem, i.e. determine the required pump pressure, with different pressure drop in each line regardless of the number of working sections of the axial divider.

To solve the "inverse" problem, i.e. determining the pressure in any of the trunk at a known pressure pump, we can express from the equation (3) for example, ΔP_1 :

$$\Delta P_1 = P_n \cdot \frac{tg(\gamma_1) + tg(\gamma_2) + tg(\gamma_3) + tg(\gamma_4)}{tg(\gamma_1)} - \frac{\Delta P_2 \cdot tg(\gamma_2) + \Delta P_3 \cdot tg(\gamma_3) + \Delta P_4 \cdot tg(\gamma_4)}{tg(\gamma_1)}. \tag{4}$$

On the basis of (4), we obtain the generalized equation of pressure drop for any i- section:

$$\Delta P_i = \left(P_n \cdot \sum_{j=1}^{k} \left(tg(\gamma_j)\right) - \sum_{j=1}^{k} \left(\Delta P_j \cdot tg(\gamma_j)\right)_{\text{при } j \neq i} \right) \Big/ tg(\gamma_i). \tag{5}$$

IOP Conf. Series: Materials Science and Engineering **134** (2016) 012039 doi:10.1088/1757-899X/134/1/012039

The shafts of the sections of the flow divider are rigidly interconnected, so the frequency of rotation and costs, passing through the sections can be determined by the following relationship:

$$Q_i = Q_n \cdot tg\left(\gamma_i\right) \Big/ \sum_{j=1}^{k} tg\left(\gamma_j\right). \qquad (7)$$

For the numerical simulation of allocation process of expenditures Q_i and pressure drops across the sections ΔP_i in the volume of the flow divider at different angles of regulation, we use the software package «MathCad v.15».Thus, for clarity and ease of analysis of results let us set the following conditions:

- all sections of the divider have the same diameter of piston d_i = 10 mm, the same diameter of the location of the axes of the plungers in the cylinder D_i = 80 mm, the same largest angle regulation $\gamma_м$= 30°;все секции делителя имеют одинаковые диаметры плунжеров d_i = 10 мм, одинаковые диаметры расположения осей плунжеров в блоке цилиндров D_i = 80 мм, одинаковые наибольшие углы регулирования$\gamma_м$ = 30°;
- the magnitude of the pressure drop ΔP_i in the output highways of the sections (see Fig. 1)depends on pressure drop in a power hydraulic motors of the motor-wheels, due to the fact that hydrostatic actuators pressure loss in the piping, usually, much less than the pressure drop in the hydraulic motor, accept: ΔP_1= 30 МПа, $\Delta P_2 = \Delta P_3 = \Delta P_4$= 5 МПа;
- pump flow Q_n = 0,5m^3/min, volumetric efficiency of the hydraulic machines is not considered.

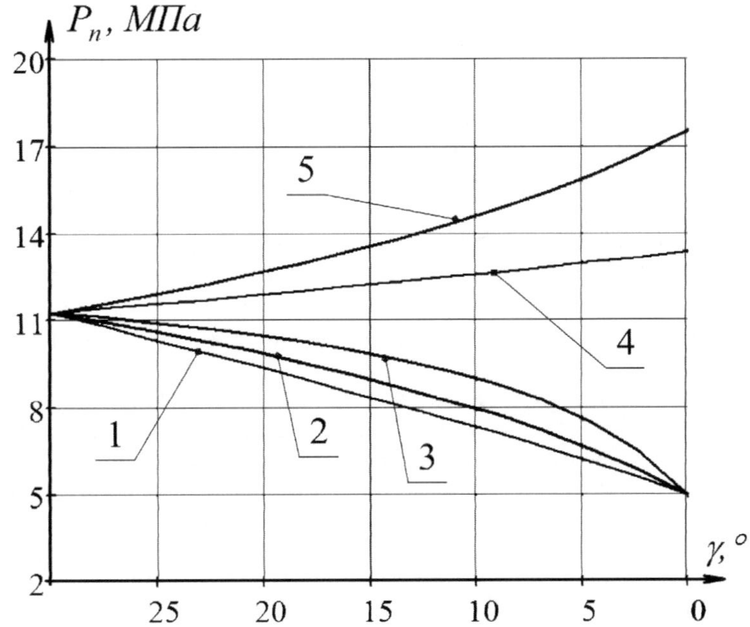

Figure 2 – Dependence of pressure P_n pump from the corners of γ adjustment in sections.
1- adjustment γ_1; 2- adjustment γ_1 *and* γ_2; 3- adjustment γ_1, γ_2 *and* γ_3; 4- adjustment γ_2; 2- adjustment γ_2 *and* γ_3;

The simulation results are presented in the form of the dependence of the pump pressure, cost through the section and possible pressure drop for any section of the divider from the corners of adjustment (see Fig. 2, 3 and 4) obtained by the equations (3), (5) and (7) respectively.

Dependencies show (see Fig. 2) that the pressure P_n in the system of hydraulic transmissions is less than the pressure in the loaded circuit ΔP_1, except when lightly loaded in three sections 2, 3 and 4 the angles of inclination of the washers is equal to 0 (not shown). Moreover, the decrease in the angle

IOP Conf. Series: Materials Science and Engineering **134** (2016) 012039 doi:10.1088/1757-899X/134/1/012039

γ1 in highly loaded sections, the pump pressure is reduced until the load of the remaining sections (curves 1, 2 and 3), and when the angle is decreased in the lightly loaded sections, the pump pressure increases (curves 4 and 5).

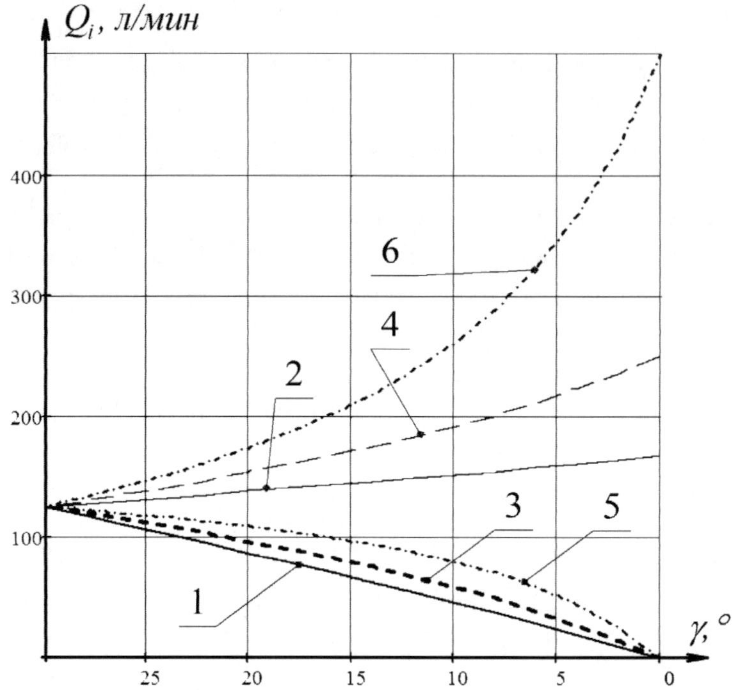

Figure 3 – Dependence of Q_i through the expenses section from the corners of γ adjustment.
1- $Q_1=f(\gamma_1)$; 2- $Q_4=f(\gamma_1)$; 3- $Q_1=f(\gamma_1,\gamma_2)$; 4- $Q_4=f(\gamma_1,\gamma_2)$; 5- $Q_1=f(\gamma_1,\gamma_2,\gamma_3)$; 6- $Q4=f(\gamma_1,\gamma_2,\gamma_3)$;

The change of the adjustment sections of the flow divider pump flow $Q_n=const$ causes a redistribution of costs through section (see Fig. 3), i.e., when the angle is decreased, the flow through the adjustable section is reduced (curves 1, 3 and 5), and using the unregulated section- increases (curves 2, 4 and 6) is proportional to the number of unregulated sections and angles of inclination of the washers.

At constant pump pressure (see Fig. 4, straight 7) reduction of the angle adjustment section increases the amount of pressure in the circuit after the flow divider adjustable sections (curves 1 and 5) and to decrease the pressure in the circuit after the flow divider unregulated sections (curves 2, 3 and 6). Curve 4 shows that a decrease in the angle control in high-load section leads to a slight increase in pressure in the low-load circuits of unregulated sections.

Thus, the use in hydrostatic transmission MM of an adjustable axial flow divider gives the ability to redistribute power flow of the working fluid. This will allow us to extend the adjustment range of traction-dynamic characteristics of MM, which will significantly improve the mobility and the controllability of the vehicle without increasing the capacity of the power plant as a whole.

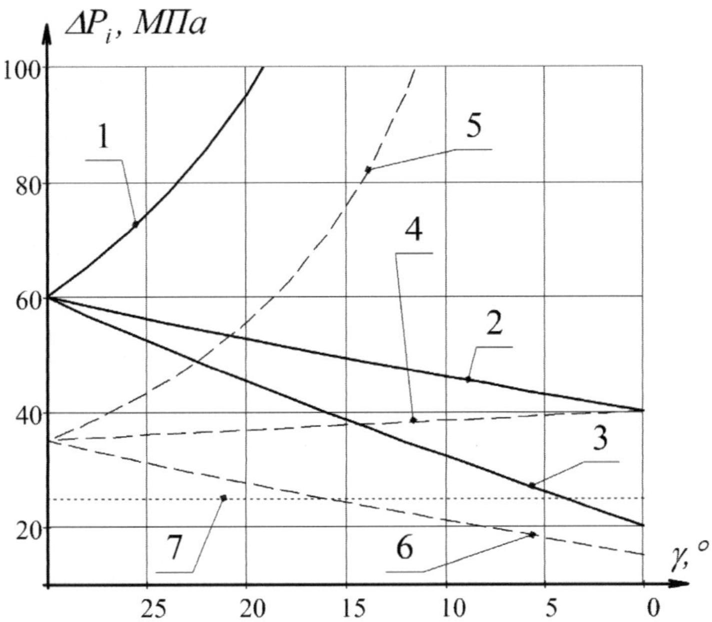

Figure 4 – Dependence of pressure drop $\Delta P_i d$ in the contours from the angles of γ adjustments in the sections of the divider.

1- $\Delta P_1=f(\gamma_1)$; 2- $\Delta P_1=f(\gamma_2)$; 3- $\Delta P_1=f(\gamma_2,\gamma_3)$;4- $\Delta P_2=f(\gamma_1)$; 5- $\Delta P_2=f(\gamma_2)$; 6- $\Delta P_2=f(\gamma_3)$ 7- P_n;

References

[1] Andreev M A and Semenov S E 2015 *Evaluation of the energy efficiency of hydraulic drive for auto wheels with different modes of regulation* Science and Education The MSTU by N E Bauman. Electronic journal 12

[2] Golubev V I, Zuev Yu Yu, Dragomirov D V 2010 *Research of characteristics of a volumetric hydraulic transmission with adjustable twin-engine unit* Herald of MEI 2 Pp 5–14

[3] Dragomirov D V, Golubev V I, Zuev Yu Yu, Fedenkov V V *Research of the hydraulic motor unit with an extended range of stepless speed regulation* // Construction and road machines, 2010. No.4. Pp. 28-31.

[4] Golubev V I, Zuev Yu Yu, Zueva E Yu 2012 *Features of work and modeling of the control characteristics of volumetric hydraulic drive with frequency and machine control* Herald of MEI. 1

[5] *The vehicle-terrain with a wheel formula 8x8*: Pat 153670 Rus. Federation. No 2014147602/11

[6] Karelin D L, Gureev V M and Mulyukin V L 2014 *Multi-circuit cooling system based on the volumetric flow divider* Automotive No 5 pp 13-15

[7] Bashta T M, Rudnev S S and Nekrasov B B 1982 *Hydraulics, hydraulic machines and hydraulic drives* M.: Engineering p 423

The calculation of a thermal field in the surface of a processed partunder the influence of a low-temperature plasma

AD Samigullin, AT Galiakbarov, AT Gabdrakhmanov, RT Galiakbarov.

Kazan Federal University, 423800, 68/19 Mira avenue, NaberezhnyeChelny, Russia

samigullin86@mail.ru

Abstract: This paper presents the results of the calculation the thermal field distribution on the surface of processed partunder the influence of a moving heat source (of the low-temperature plasma).

1. Introduction:

The most widespread task in the treatment with low-temperature plasma is to obtain specified characteristics (depth, width, heat treatment, etc.)depending on parameters of the heat source, i.e. the current, the arc voltage, the conveying speed, and other parameters of the regime. Moreover, as a rule, a necessity appears to receive the heat treatment with certain physical and mechanical properties [1].

2. Purpose:

In selecting of thetreatment conditions is necessary to analyze of the nature heat distribution in detail, the evaluation of the characteristics of the thermal field is determined by the quality of thedetailreceived.A method for determining the interconnection of parameters in complex processes is a mathematical modeling. This method is based on the fact that the study is not subject to the phenomenon of interest to us or a process, the study is subject a mathematical model - the image of a real object or display, building by means of mathematical relationships that establish a link between the defining characteristics of the object. Another important factor is the relatively low cost of the computational experiment as compared to physical.

3. Simulation:

Due to high temperatures and flows with a high concentration power the detailed experimental study and visualization of the temperature field distribution on the surface ofprocessed part is associated with considerable difficulties.The visualization of the temperature distribution on the surface of processed part is necessary for quality control, and the heat treatment process. In the MathCadwas created by the distribution model of the temperature field on the surface of processed part by moving the heat source.

Initial data for this model is the thermal characteristics and size of the material, the heat treatment parameters like the current, the voltage and the efficiency of the plant, which in turn affect the capacity of the heat source q, the processing speed.

The heating period in a moving coordinate system is not only a functionthe coordinates but also of the time. The temperature field in the heating stage is described by the formula [3]

$$\Delta T(x, y, t_H) = \frac{q}{4\pi\lambda\delta} e^{-\frac{vx}{2a}} \int_0^{t_H} \frac{1}{t} e^{-\left(\frac{v^2 t}{4a} + \frac{r(x,y)^2}{4at}\right)} dt \quad (2)$$

whereλ - the thermal conductivity;
tH - the duration of the heat source;
q - the power the heat source;
v –the conveying speed of the heat source;
a - the thermal diffusivity.
The results of the model of the temperature distribution on the surface of processed part shown in Figure 1.

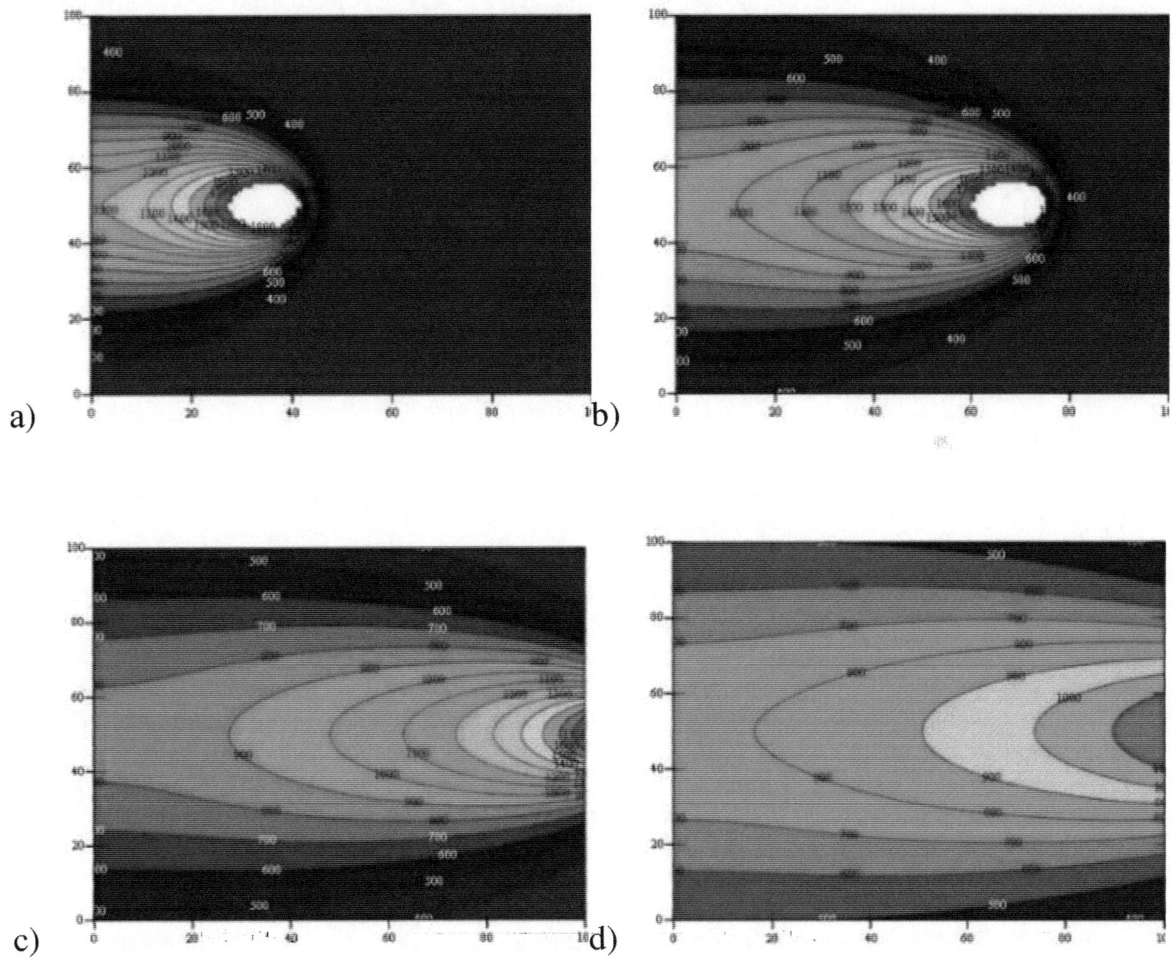

Figure1. Computer modeling of the temperature field distribution on the surface of processed part in the program Mathcad. a - after 1 c, i - after 2 c, c - after 3 c, d - after 4 c.

According to the calculated data revealed that on the surface of the metal steel 40formed a region in which the temperature is about 2000 K. Then conducted field tests confirm the theoretical calculations (Fig. 2)in particular band is observed with melting the metal from the plasma. Having a graphical representation of the temperature field on the metal surface of the heat affected zone is defined and the structure of the metal after influence of the heat source.

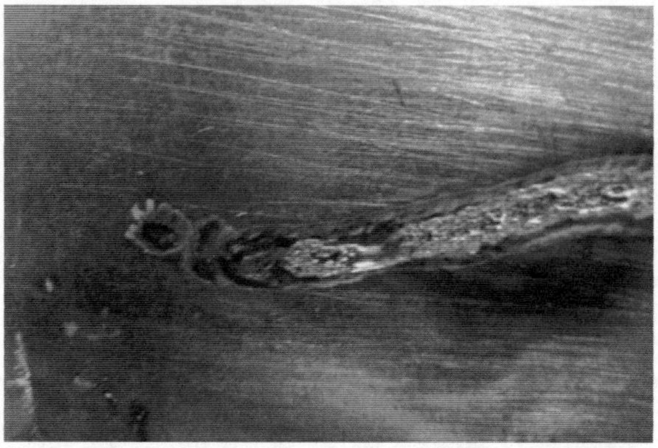

Figure 2. Traces of melting on the surface of processed part after the low-temperature plasma treatment.

4. Conclusion:

On the basis of the above mentioned model, after receiving the picture of the distribution of the temperature field on the surface of processed part have the opportunity to choose the desired mode of processing, without conducting field experiments to determine the optimal treatment regime.

References

[1] *The calculation of of temperature fields during welding and surfacing*: Textbook / AY Medvedev; Ufimsk state aviation tehn Univ - Ufa 2009 142 p

[2] Deacons V V 2000 MathCAD 2000: *Training. Course* SPb .: Publishing House, Peter 592 p.

[3] VN Volochenko VM Yampolsky VA Vinokourov etc .1988 The theory of welding processes: *Proc. for high schools on special. "Equipment and technology of welding production"* M.: Higher HQ 559 pp.

[4] Samigullin A D 2015 *The study of gas-dynamic and energetic parameters of the bulk plasma generator*/ Socio-economic and technical systems: research design optimization Number *4 (67)* pp 30-38

[5] AT Gabdrakhmanov, Israfilov IH, Galiakbarov AT, AD Samigullin 2013 *The study of motion of an electric arc in a pulsed plasma generator* IRTC IMTOM 2013 Forum Part 2 pp 37-40

[6] Israfilov I H Galiakbarov A T Bashmakov D A Gabdrahmanov A T Samigullin A D 2014 *Pulse* plasma surface thermostrengthening of machine *parts IOP Conference Series: Materials Science and Engineering* Volume ***69*** *012037*

[7] *I K Israfilov, A T Galiakbarov, D I Israfilov 2014* The analysis of the bottom electrode thermal streams during work process in a melting furnace IOP Conference Series: Materials Science and Engineering **69** 012015

[8] A D Samigullin, A T Galiakbarov and R T Galiakbarov 2016 Study of the petroleum schedules thermal cleaning process from asphalt, ressin and paraffin deposits using low- temperature plasma Journal of Physics: Conference Series **669** 012017

[9] Israphilov I H, Bashmakov D A, Galiakbarov A T, Mandrik P A 2014 The mathematical description of the thermal combined laser and plasma impact on a material *Contemporary Engineering Sciences* V 12. pp 569-576

[10] I H Israphilov, D I Israphilov, D A Bashmakov, A T Galiakbarov, A D Samigullin 2015 *Calculation of thermal processes in bottom electrode* Contemporary Engineering Sciences Vol 8 p 13-20

[8] Israfilov Z K and Kashapov N F 1991 *Steadying the instability of a glowing discharge in a longitudinal air stream.* Journal of Engineering Physics60 364–368
[9] Fayrushin I, Kashapov N and Dautov I 2014 *Calculation of distribution of potential and electron concentration in the dust-electron thermal plasma with the axial geometry particles* //IOP Conference Series: Materials Science and Engineering J. Phys.: Conf. Ser.567 012009
[11] Gavrilova V A, Fazlyyyakhmatov M G and Kashapov N F 2013 *Protective matching polymer powder coating of piezoelectric element* // Contemporary Engineering Sciences.J. Phys.: Conf. Ser.479 012010

IOP Conf. Series: Materials Science and Engineering **134** (2016) 012041 doi:10.1088/1757-899X/134/1/012041

Selection of equipment for machining processing of parts using NX and TEAMCENTER programs

P M Krastyaninov and R M Khusainov

The Department "Design and technological support of machine-building production"
Kazan Federal University, Naberezhnye Chelny Institute
Pr. Mira 68/19, Naberezhnye Chelny, Republic of Tatarstan, Russia, 423812,

rmh@inbox.ru

Annotation: The article describes the current methods for selecting of process equipment for machining of parts on metal cutting machines. The authors suggest the method of selection of equipment based on the use of NX and Teamcenter software as well as the recommendations to improve the program's database of Teamcenter. The offered method allows to reduce the terms of technological preparation of production due to the simultaneous selection of equipment, tools, equipment and the formation of the control program.

Currently, we have the most common method of selecting of process equipment for machining of parts, when the technology chooses hardware, relying only on his experience. The disadvantages of this method are the requirements for the qualification of a technologist and a great time for search of details about this available equipment. A literature review showed that other methods have been offered, such as the use of a data bank [1] and the so-called "modular technology" [2]. The disadvantages of these methods is the excessive formalization of the procedure, and as a result, you can not answer the problem of nonstandard.

A method based on the use of simulation processing in the module of NX CAM system and

IOP Conf. Series: Materials Science and Engineering **134** (2016) 012041 doi:10.1088/1757-899X/134/1/012041

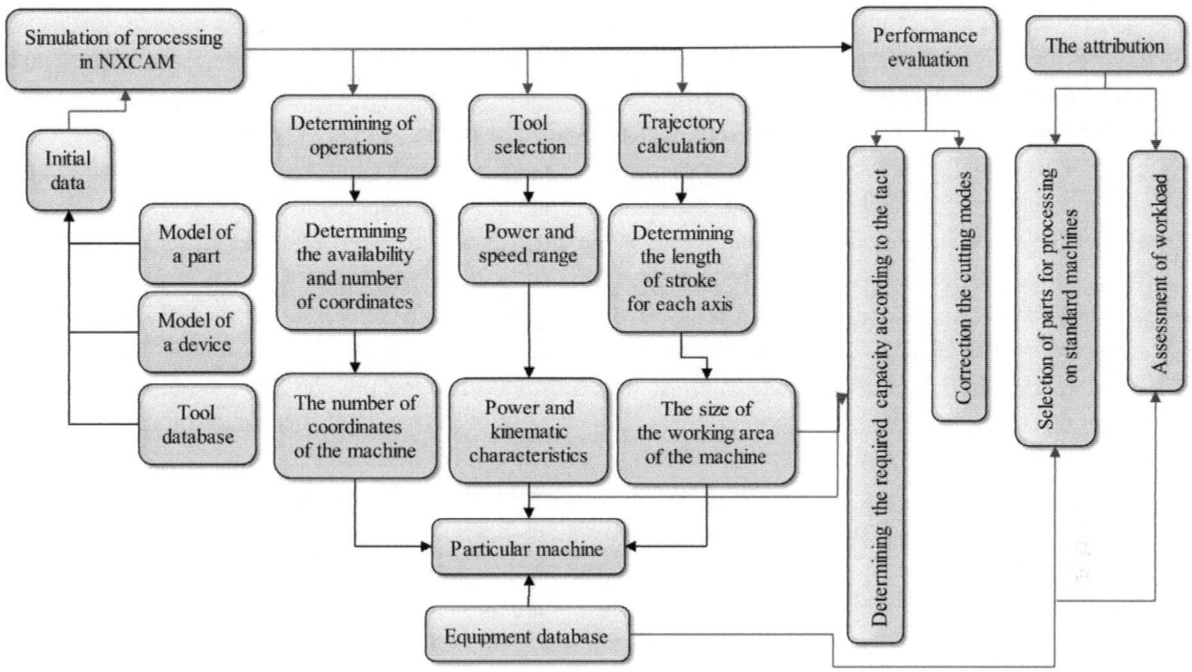

Figure 1 - The order of selection of the equipment

The relationship between the parameters of a part the TA and the parameters of the machine was determined in order to use this technique.

Table 1 - Parameters of a part- description of the equipment.

Parameters of a part	Equipment perrformance
Shape of a part	The type of equipment movements (number and types), the type of machine adjustment
Overall dimensions	Working space
The number and kinds of processed surfaces	Movements (number and types), the amount of bearing instrument and its nomenclature
Workpiece material	Cutting, power drives
Accuracy of part manufacturing	Accuracy of machine
Roughness	Cutting conditions
Tact and the release program	Productivity

The initial data are three-dimensional models of parts and blanks, and later the models of equipment and tools [3].

At the first stage the operation is created in the module NX CAM. As a result of this step the number of required coordinates is determined while the surface treatment as well as the required length and diameter of the tool, the range of surface treatment with a single install. These results provide baseline data for further procedures.

For further calculations it is necessary to choose thetool. Its length and diameter are selected from the previous calculation, the necessary tool is inserted from the NX library or from public sources, or from Teamcenter, with the possible creation of an assembly tool [4].

Further, the cutting conditions are determined according to the recommendations of the manufacturer tool. The main criteria necessary for choosing the cutting modes are a material part, tool and cutting method [5].. And it is possible to choose the modes from the NX library. Then we find the required power of cutting, and paste it into MS Excel file titled "Power of of Cutting". The result of this phase is the cutting power by which the capacity of main drive is determined, the required spindle speed and the required feed, which serves as a guideline for the selection of the machine with the appropriate parameters.

The next step - determining the length of stroke. The stroke length is determined either from the tool control program or by the simulation of the machine. As a result of this stage the length of the stroke as well as the required size of the working area of the machine are determined.

Thereafter, the files are attached to an item in the items database Teamcenter as attributes.

Currently, the application Classificator is used to describe all the resources available in the enterprise. The search of tools is also carried out in this application. Technological equipment is described in two tabs: "Resource Management" and "Hardware".

The tab "Resource Management" describes the following resources: Fastening devices, Machines and Devices, Automated robots and Welding tongs tools.

In the Hardware tab, classification is carried out on several types: By type (metal cutting, automatic transfer lines, assembly, etc.), the production (Foundry, Engine plant, etc.), model (metal-cutting equipment, automated production lines, assembly etc.).

In the program NX machine tool is inserted from the tab "Resource Management". In the search for tools the received Excel files are being opened and the desired values are input in the appropriate box. The system finds a certain number of machines and it is still necessary to determine whether the selected machine suits by the tool performance. For this it is proposed to introduce the "load factor Ks" to identify opportunities for the production of new parts. It is also recommended to enter the column "Plant" and "Inventory number" in the tab "Resources Management" to exclude transitions between the Applications tab and reduce the time to choose machines. If Ks is above 0.85, the search is on (on another inventory number or other plant). Only then we can finally assign the equipment to a particular item.

When the simulation program is running the run time is displayed, and through Ks we assess, whether the machine tool is able to handle this item with the specified performance.

Thus, the proposed method allows to simultaneously perform the following tasks: to shape UP, tool selection, calculation of cutting conditions and selection of equipment. As a result, it reduces the time and increases the efficiency of CCI of choosing equipment [6]..

Bibliography:

[1] Kalinin M V 2006 *Investigation of the effect of design and technological parameters of prismatic parts to choose from multi-drilling, milling and boring machines Dis. cand. tehn. sciences* (Moscow: "ENIMS") pp 153 - 88

[2] Kryukov A V, Arzybaev A M and Sakharov AV 2012 *Modular technology as the basis for the formation of groups of machine problems in the operational planning of production technology Mechanical Engineering* (Moscow: Mechanical Engineering) №**2** pp 56-9

[3] Balabanov I P, Balabanova O N and Groshev A V 2015 *Formation of initial data of the workpiece batch in simulation modeling precision forming* IOP Conf. Ser.: Mater. Sci. Eng. **86** 012011

[4] Khisamutdinov R M and Khisamutdinov M R 2014 *Automation system goals for the creation and operation of the tool* IOP Conf. Ser.: Mater. Sci. Eng. **69** 012021

[5] Ziyatdinov R R 2014 *Method of calculating of optimal grading modes with regard to forming of grinding surface roughness* IOP Conf. Ser.: Mater. Sci. Eng **69** 012044

[6] Kharisova A R and Puryaev A S 2014 *Competitiveness assessment of engineering products* IOP Conf. Ser.: Mater. Sci. Eng **69** 012020

IOP Conf. Series: Materials Science and Engineering **134** (2016) 012042 doi:10.1088/1757-899X/134/1/012042

Methods of assessing the dynamic stability of the cutting process using UNIGRAPHICS NX

R M Khusainov and I F Sharafutdinov

Department "Design and technological support of machine-building production" Kazan Federal University, Naberezhnye Chelny Institute
Pr. Mira 68/19, Naberezhnye Chelny, Republic of Tatarstan, Russia, 423812,

rmh@inbox.ru

Annotation. The article deals with the determination of the stability of the cutting process at the Nyquist criterion. Attention is paid to the use of CAE systems to solve this problem. The result of this method is to improve the reliability of technological operations.

In today's production of metal, when high-speed machining becomes widespread [1], the task of providing the dynamic stability of machine tools comes to the fore, specifically the ability of the dynamic system of the machine to resist the emergence of self-oscillation. It depends on the static and dynamic characteristics of all the elements involved in technological system [2], and the parameters of the cutting mode. The most commonly used criterion for assessing the stability of a dynamical system is the frequency Nyquist criterion, according to which the dynamic system will be stable, if the graph of the amplitude-phase frequency characteristic (AFC) does not include the point with coordinates [-1, i0] on the real axis.

$$W_{dc}(i\omega) = W_{ees}(i\omega) \cdot W_{cp}(i\omega) \qquad (1),$$

where $W_{dc}(i\omega)$ - transfer function of the dynamic system, $W_{ees}(i\omega)$ - transfer function of the equivalent elastic system of the machine together with tool, an instrument and the workpiece, $W_{cp}(i\omega)$ - transfer function of the cutting process.

Nowadays, with the widespread technology of "digital production", it is possible and appropriate it is possible and appropriate to make a calculation of these functions with the use of three-dimensional modeling and finite element analysis [3].

One of the most common systems for solving problems of "digital production" is UNIGRAPHICS NX. One way to obtain an equivalent response of the elastic system in this software is as follows:

1. The preparation of modular three-dimensional model of the machine with a device, tool and workpiece, and technological elements of the system should be the ones that participate in the processing of the part.
2. The dynamic calculation of a prepared model is made in NX Advanced Simulation . module. Setting solution is a frequency analysis of the direct method (SOL 108 - Direct Frequency Response).

To build a APFC it is important to note the following: in the setting step, in the *Request output* parameter you must move the tab to put option IMAG, which allows us to derive the real and imaginary part of the transfer function, allowing you deduce the module and phase of the transfer function. In both cases calculated data will is sufficient to construct APFC.

3. At the end of the calculation it is necessary to select the assembly *NastranOP2 Files - Sort2 Results* in the *XY Navigator*, upload a results file, from the list from the list *Displatsement* select a component of movement, which strongly affects the accuracy and the change of cutting parameters (in the case of turning - along the radius of the workpiece) [4].
4. The transfer function of the cutting process can be obtained from the following relations:

$$W_{cp}(i\omega) = \frac{K_p}{1+T_p^2\omega^2} - i\frac{K_p T_p \omega}{1+T_p^2\omega^2} \quad (2),$$

where Kp - cutting rigidity; ω - angular frequency; i - imaginary unit; Tp - time constant of the chip. These parameters are determined by the cutting conditions according to [5].

Therefore data required to calculate the transfer functions can be calculated in MicrosoftExcel or a mathematical packages. It is important that these data were calculated with the same frequency as in n. 2.

5. The resulting transfer function of the dynamic system is determined using the following relations:

Module of the transfer function:

$$A_{dc} = A_{ees} \cdot A_{cp} \quad (3),$$

where A_{dc} – the module of the dynamic system, A_{ees}- the module of the transfer function of the equivalent elastic system, the A_{cp} - the module of the transfer function of the cutting process.

Phase:

$$\varphi_{dc} = \varphi_{ees} + \varphi_{cp} \quad (4),$$

Where φ_{dc} – phase of the dynamic system, φ_{ees}- phase of the transfer function of the equivalent elastic system, φ_{cp} - phase of the transfer function of the cutting process.

So you can build a dynamic response of the system and evaluate its stability according to the Nyquist criterion.

Application of this technique allows to evaluate the possibility of the technological system to ensure quality, and performance with the release of a given product. Application of three-dimensional model and design applications allows you to find the weak link of the technological system and carry out targeted operational measures in order to ensure the production of the given parameters [6].

IOP Conf. Series: Materials Science and Engineering **134** (2016) 012042 doi:10.1088/1757-899X/134/1/012042

Bibliography:

[1] Khisamutdinov R M and Khisamutdinov M R 2014 *Automation system goals for the creation and operation of the tool* IOP Conf. Ser.: Mater. Sci.Eng. **69** 012021

[2] Gumerov I F, Shafigullin L N, Vakhitova S M and Shaekhova I F 2014 *Noise absorbing composite materials applied in domestic trucks* IOP Conf. Ser.: Mater. Sci.Eng. **69** 012011

[3] Sibgatullin K E and Sibgatullin E S 2014 *The determining of the coefficient of safety of bearing ability of anisotropic bars in the general case of their complex resistance* IOP Conf. Ser.: Mater. Sci.Eng. **69** 012041

[4] Goncharov P S, Artamonov I A, Halitov T F , Denisihin S V and Sotnik D E 2012 *NX Advanced Simulation.* Engineering Analysis (Moscow: DMK Press) p 315

[5] Kudinov V A 1967 *Machines Dynamics* (Moscow: Mechanical Engineering) p 213

[6] Shibakov V G, Pankratov D L and Shibakov R V 2014 *Connection between machines durability during operation and the complex of physic-mechanical properties, formed during the production of critical parts* IOP Conf. Ser.: Mater. Sci.Eng. **69** 012040

IOP Conf. Series: Materials Science and Engineering **134** (2016) 012043 doi:10.1088/1757-899X/134/1/012043

The algorithm of the expert system of APCS (automated process control systems) tests

D T Yakupov, G I Nabiullina and I S Sabirov

Kazan Federal University, Naberezhnye Chelny Institute
Pr. Mira 68/19, Naberezhnye Chelny, Republic of Tatarstan, Russia, 423812

yaqup@mail.ru

Abstract: description of a general algorithm of the expert system of APCS tests is presented.

Starting of automated process control systems (APCS) on-site involve risks of expensive equipment failure because of critical transient violations of manufacturing processes due to the need of unplanned shutdowns of the equipment, there may be situations of increased danger to personnel and the environment. According to [1], by modeling the production processes at the stage of design of APCS it's possible to avoid negative consequences of the commissioning process. This work is carried out not on the technology object (TO), but on the virtual one, created by the mathematical and computer modeling tools, using the hardware-software test complex.

Actions for testing the APCS have the following objectives: troubleshooting hardware control systems, detecting algorithm control of technological units mismatch the required parameters, simulation of operating and emergency operating conditions of TO.

Intelligent add-in is designed to find the causes of any anomalies and issue recommendations on removal of factors of unacceptable technology object work. The anomaly here is the set of test parameters, that doesn't meet the given constraints [2].

Figure 1 shows a block diagram of a general algorithm of the expert system of APCS tests. Operation of the system begins with the selection by a user of a test group. Testing of a functionally complete control cabinet is divided into three groups:

- checking the appropriate signals to the input/output (correct connection of input/output signals to a controller equipment is determined);
- testing of the control system in manual mode (correct operation of the system in the management of technological units in manual mode is measured) ;
- testing of the control system in automatic mode (the validity of the control system reaction on the simulated state of the technology object is determined).

Each group corresponds to the test frame, an example of which is formed after the selection of the group.

IOP Conf. Series: Materials Science and Engineering **134** (2016) 012043 doi:10.1088/1757-899X/134/1/012043

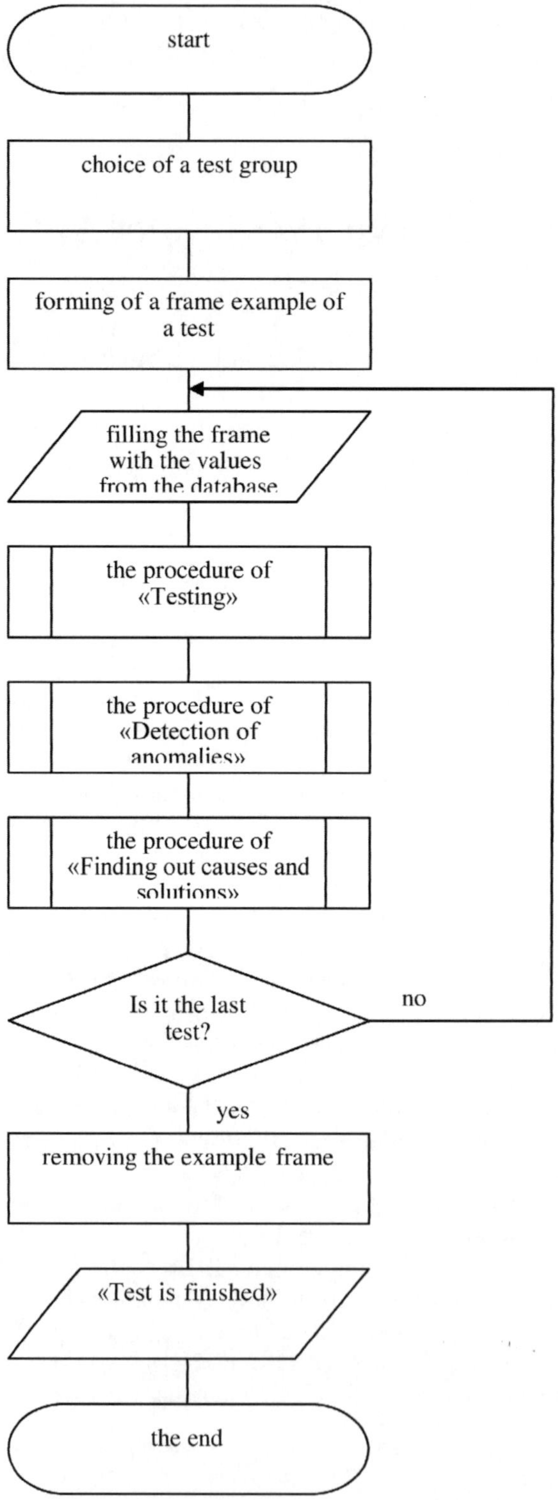

Figure 1 - a block diagram of algorithm of the expert system of APCS tests.

IOP Conf. Series: Materials Science and Engineering **134** (2016) 012043 doi:10.1088/1757-899X/134/1/012043

Test procedure is performed. Each group has its own version of the test. When checking the conformity of the input signals of the system under test, the signals of the respective output channels of testing system are determined and set, information on the status of input is read through interface connection. When checking the conformity of the signals at the outputs of the test control output status read via an interface, also the relevant inputs of a test complex are interviewed.

In testing of the control system in manual mode, in accordance with the data from the database, dialog boxes that indicate the required action for the setter are generated [3]. Feedback is implemented by reading signals from the inputs of the complex and the connection interface. When testing a control system in an automatic mode in accordance with a predetermined sequence, specific signals are formed, which are sent to the control system. Feedback is implemented by reading signals from the input of the complex and the connection interface. On the results of tests slots are filled with the values of previously generated frames. The procedure of anomaly detection also operates with different groups of tests. When checking the conformity of input/output values read from the controller system under test, and the values pre-set at the stage of parameterization are compared.

If the value of the real signal is considerably higher (lower) than the set value the anomaly "high (low) level signal» is generated. In the absence of a signal on the current channel the anomaly "No signal" is formed. Then there is the validation of indication.

Work testing in manual and automatic modes occurs in three modes: the launch of a mechanism, stopping the mechanism, regulation of a parameter. If there is no signal that the mechanism has launched (stopped), connection with the equipment is checked: if there is no connection the anomaly "There is no connection with the equipment" is generated, otherwise - the anomaly " Mechanism has not launched (has not stopped)" is generated.

When the position of the manual parameter setter is changed the control of this parameter is checked. If there is no signal that the parameter is adjusted connection with the equipment is checked: if there is no connection, there is the generation of anomaly "There is no connection with the equipment", otherwise - the anomaly "Parameter cannot be adjusted" is generated.

When adjusting the parameters the quality of adjustment is automatically checked. If the parameter adjustment does not satisfy the constraints the anomaly "Poor transient response" is generated. In automatic mode of the control system work also the system response to simulated exposure to a pre-set control algorithm is checked. If the system response does not match the algorithm a corresponding anomaly is generated. If the reaction of the system corresponds to the algorithm, the state of the virtual technology object is examined: performance indicators of nodes and mechanisms are calculated using a mathematical model of the control object. If calculations indicate damage to any nodes the anomaly "Emergency operation of the equipment" is generated. Then the work condition of the devices and circuit protection mechanisms is checked. If the protection device should have worked, but it hasn't, the anomaly "The protection device hasn't worked" is generated. In case where the protection device should not have worked, but it has worked, the anomaly "The protection device has worked» is generated.

If anomalies are not detected and it's not the last test, a new test cycle begins. If the anomaly is detected, search for the causes of anomalies in the base of production rules is made. If there are multiple causes, it is necessary to resolve the conflict by choosing the most probable cause of the current anomaly.

The detected causes are entered in the appropriate frame, and then the search for solutions of the current anomaly starts. If there are multiple solutions, you must resolve the conflict by choosing the most probable solution of the anomaly. The solution is entered in the appropriate frame. After determining the cause and finding a solution, the formation of explanation and displaying it on the screen takes place.

In case when the solution is recognized invalid the transition to the next solution occurs, in the absence of a solution - to the analysis of the next cause. In the absence of causes and solutions of an anomaly a message on the impossibility of solving the current anomaly is displayed. If the solution is

IOP Conf. Series: Materials Science and Engineering **134** (2016) 012043 doi:10.1088/1757-899X/134/1/012043

recognized true the increase of the significance factor of the current production rules, which led to the solution, occurs.

If the test is the final one in the group, the frame example is removed, and a message about the completion of tests is displayed. Application of the expert system in the tests would let reduce the number of errors in the development of control systems; reduce the time to identify and eliminate the causes of malfunction of the system; reduce the probability of failure of expensive equipment.

Bibliography:

[1] Denisenko V V 2009 *Computer control of a technology process, experiment, equipment* (Moscow: hotline –Telecom) p 232

[2] Yakupov D T, Sarvarov I I and Sabirov I S 2015 *Knowledge representation in intelligent systems testing APCS*. Actual problems of energy conservation and efficiency in technical systems: theses of the 2nd International Conference with the elements of scientific school (Tambov: Pershin R.V. publishing) p 207

[3] Klochkova K V, Petrovich S V, Simonova L A and Yusupov L R 2015 *Development of methodology for controlling the parameters of TP* IOP Conference Series: Materials Science and Engineering p 86

IOP Conf. Series: Materials Science and Engineering **134** (2016) 012044 doi:10.1088/1757-899X/134/1/012044

Improvements in the process of boss bar upset forging into a horizontal forging machine with the aim of joint knuckle forging quality improvement

D L Pankratov[1], R S Nizamov[1] and I Zh Kharisov[2]

[1] Kazan Federal University, Naberezhnye Chelny Institute
Pr. Mira 68/19, Naberezhnye Chelny, Republic of Tatarstan, Russia, 423812,
pankratovdl@gmail.com
[2] Open Joint Stock Company "KAMAZ" Forging plant, Naberezhnye Chelny
Avtozavodsky Avenue 2, 423827, the Republic of Tatarstan, Russia,
nizamovrs@yandex.ru

Abstract. A new technique for tapered composing transition shaping has been put forward in the process of upset forging with the use of an experimental tool. The results of the upset forging process with the use of a new composing transition has been computer simulated.

In the process of group 1 subgroup 1 (boss bar forged pieces) upset forging into a horizontal forging machine we recognize upset forging with no bending and with the restricted bending of an upset blank. The upset blank stability is determined by upset length ratio $\psi = l_{upset\,forging}/d$. The value for upset forging with no bending equals $\psi \leq 2{,}7$. Under the circumstances of this ratio overriding the upset boss bar forming is going on with the restricted bending [1]. In the process of high-carbon and alloyed steel upset blanks forging their tapered punches have critical upset length ratio being matched by spatial bending equal to $\psi_{space\,curvature} \leq 5{,}0 \div 5{,}1$, but in case of $\psi \geq 6{,}8 \div 7{,}3$ spatial concavity does not go under smoothing that in its turn hardens the process of cavity filling (fig. 1, *a*) and results in one-sided lateral burrs (fig. 1, *б*).

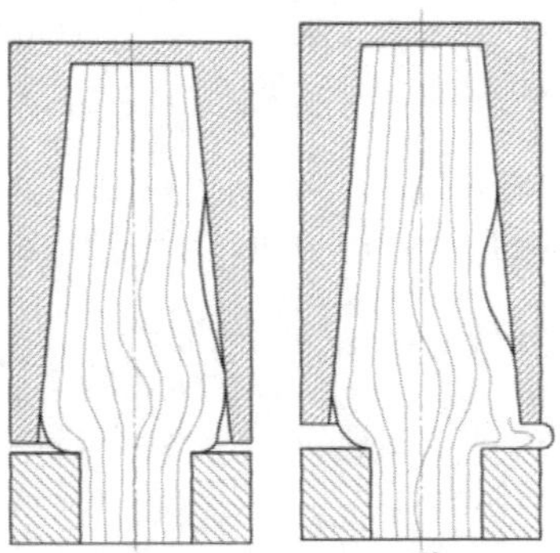

Fig.1. Shape deflection while upset forging in case of $\psi \geq \psi_{space\,curvature}$

It is suggested to use a selector punch with a cylindrical part along the full length of an upset blank running over 10d, if $\psi \geq 10$ (fig. 2) [1].

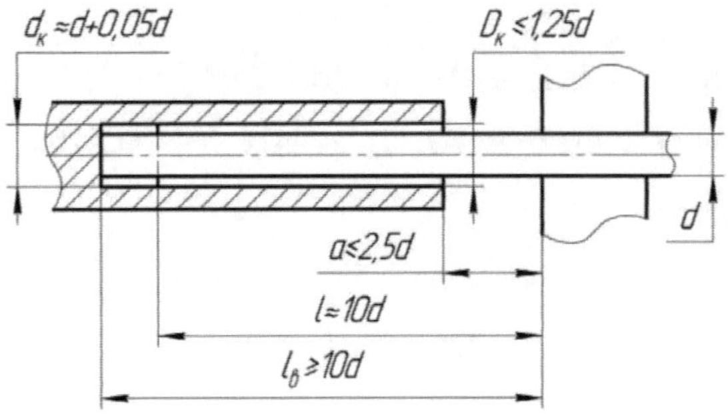

Fig. 2. Limiting values of boss dimensions and upset forging process in a tapered composing pass, if $\psi \geq 10$.

Generally, if $\psi \geq \psi_д$(allowable upset length ratio) at a bending phase on arbitrary plane XOY, point data of $0{,}25l_x$ и $0{,}75l_x$ are bending points, while point data of $0{,}2\pi l_x$ ($\approx 0{,}63l_x$) correlate with initial contact of the upset blank convex surface and tapered cavity wall. An upset blank is being deformed after coming into contact with a face of fill under side upthrust from fill wall conditions (fig.3) [1].

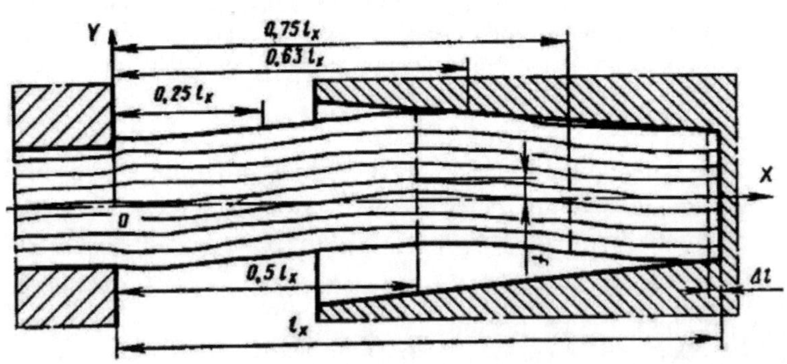

Fig.3. Bending in a plane.

From this viewpoint the example described by A.N. Bryukhanov is interesting [3]. There is a cone generatrices slope of less than 2° to upset blank axis in the first composing transitions. Beeing so minor the tapering can't eliminate one-side burr in a slot. That's why such slot transitions contemplate extended tapering along the section 10mm long. Thus, taking into account bending in a plane (fig.3) we can reason that grade reduction of a cone major section involves earlier upthrusting from a pass wall. However it can extend blank base diameter under the circumstances of volume conservation in transitions, that in actual practice results in increasing dimensions of forging devices, but it may be unacceptable in case of equipment dimensions limitation.

Production process of joint cam forging at Open Joint Stock Company "KAMAZ" Forging plant comprises 3 composing upsetting and the fourth final transition on the basis of the blank made of steel 20X2H4A Ø50. In practice the technology under discussion has an essential drawback – repeatable defects of unfilled die bed contour and burr punching in the detailed section (fig.4). These drawbacks are reflected in a completed part. One solution of the infilling problem proposes to change the 2nd and the 3rd composing transition geometry. In the study by I.V. Grezina [2], a cylindrical section in tapered punches while composing upset forging with the result of approximately correct geometry is reported.

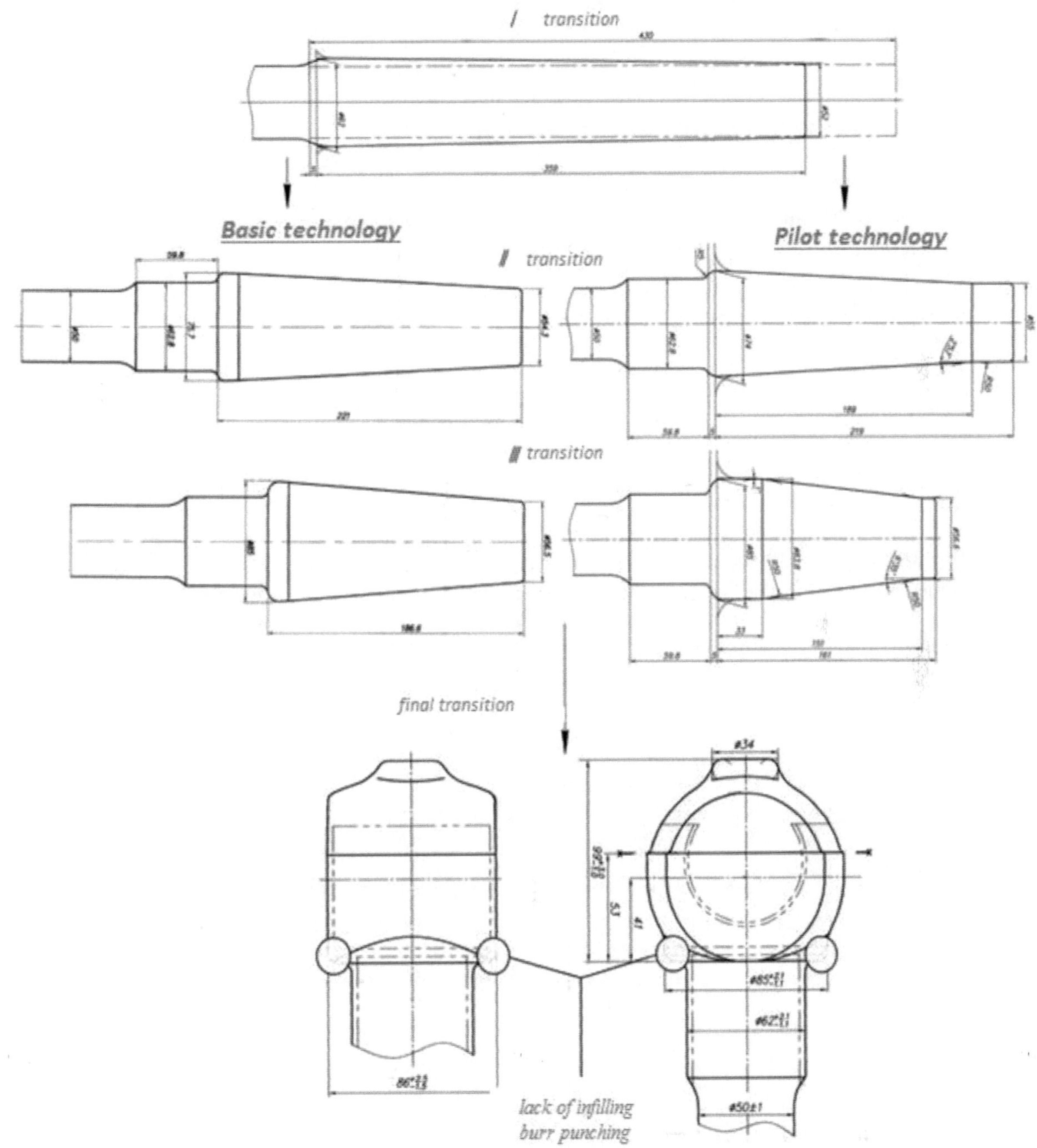

Fig.4. Composing transition forming technique with a circular section on the 2nd and the 3rd transitions and variable cone on the 3rd one.

As far as strength of materials is concerned the upset blank stability in the process of upset forging into a horizontal forging machine can be thought of as the problem of the bar stability with an overhung end and a joint one [4], whereas critical value of a compressive axial force is inversely as the bar length up to a joint end $P_{кр} = \frac{\pi^2 EJ}{(\mu l)^2}$ (fig.5, *a, б*). The usage of a circular section allows to displace a joint holdfast as much as a land portion, according to it the bar system stability increases and bending shape changes in the process of deforming – bending point data of l_x displaces (fig.3). A detailed solution to the problem of upset blank stability in a composing transition while the process of alike upset forging is given in the studies by O.S. Zhelezkov [6] and F.F. Gatin [7].

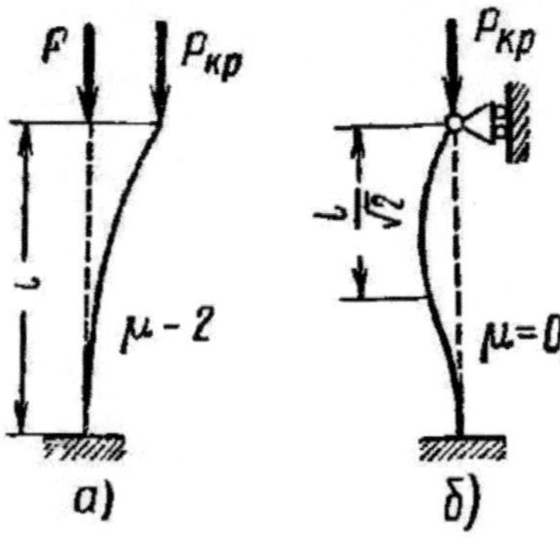

Fig.5. Compressed bars stability

Considering the equality of transition volume with the existing and pilot processes, actual values of circular sections on the 2nd and the 3rd transitions equal to ~ 8.35% and about 4.5% of the upset length respectively. The configuration of the final composing transition with a variable forming cone should be determined by the volume equality with the current technology, and a priority of metal setting in a hard-to-infill part (typically, it is a forged piece cross-section with a maximum area). In contrast to the technique by A.N. Bryukhanov [3] composing transition base slope is reduced to 1 ° and the main section slope is increased to 6 ° 35 '. This technique allows reducing the overall length of composing transition without changing the base diameter, which in its turn causes upsetting stability of a subsequent final transition to increase, i.e. coefficient ψ reduction. In the process of a forged piece upsetting, due to the originally used technology we make use of the technique by A.S. Somkina [5], which involves a clamping portion and a matrix-profiling radius changing on the transitions following the first one.

The computer simulation in the software package QForm3DV 8.0.4 for a new pilot technology of upsetting has showed the forming of the 2nd and the 3rd composing transitions' configuration required for the subsequent punching and has helped to confirm the predicted result of the uniform metal distribution increasing while the final transition upsetting (fig.6) [8].

IOP Conf. Series: Materials Science and Engineering **134** (2016) 012044 doi:10.1088/1757-899X/134/1/012044

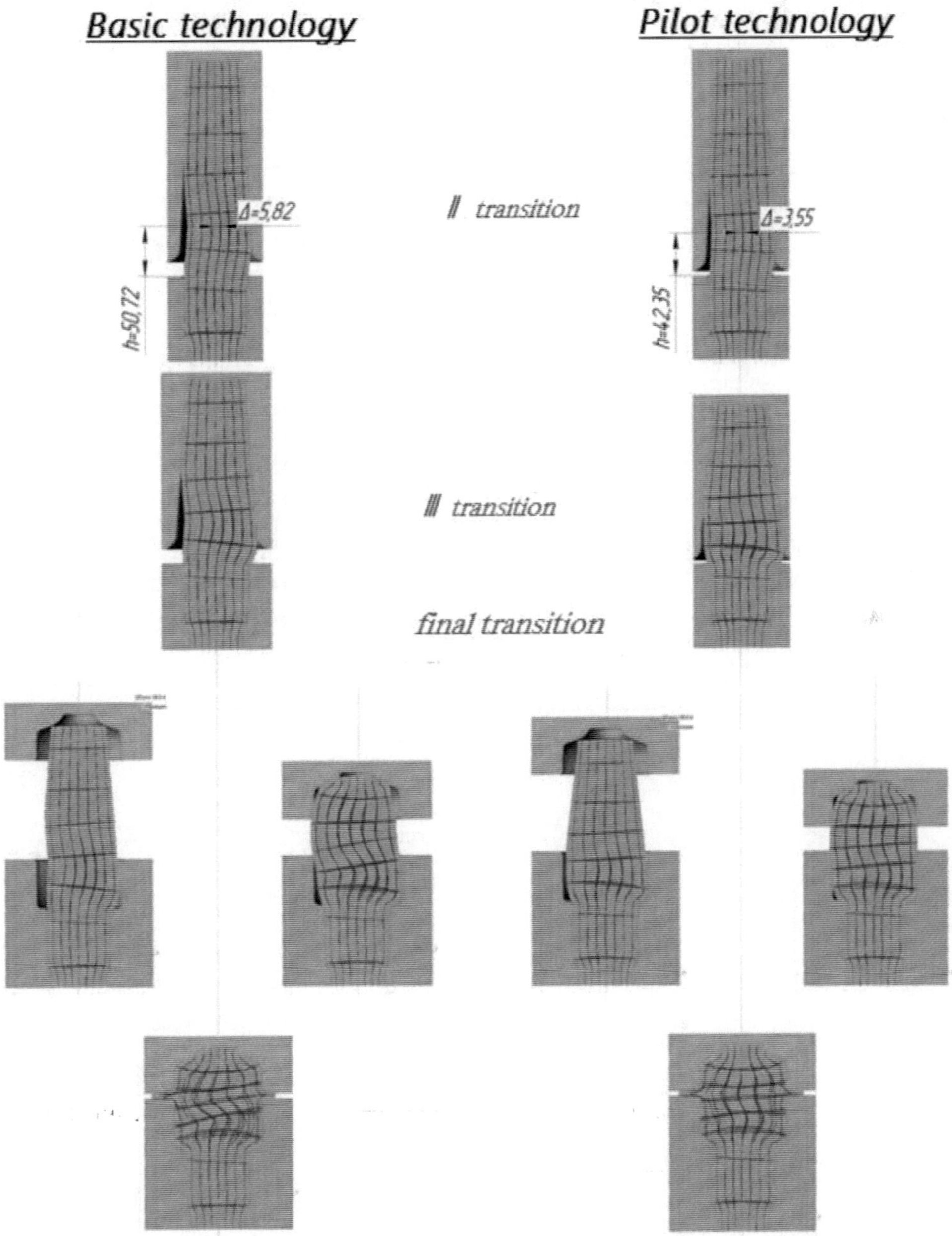

Fig.6. Computer simulating of joint cam forged piece upsetting

The comparison of the 2nd composing transition computer simulation has shown that circular section insertion with relative height of 8.35% from upset length ratio allows to reduce relative height

IOP Conf. Series: Materials Science and Engineering **134** (2016) 012044 doi:10.1088/1757-899X/134/1/012044

h of recessed undercut formation for ~17% and bending of the axis Δ for ~39% (fig.6). The final transition upsetting confirms risk minimization of such drawbacks as lack of infilling, burr punching and chuck in a completed part.

Conclusions.

The usage of final composing transition pilot technology with a variable forming cone of an element of cone along with a cylindrical section on the top of the 2nd and the 3rd composing transitions has allowed to improve uniformity of forming while upsetting and to increase completed part quality. The given solution can be used for devices engineering in the process of forged pieces upsetting group I – boss bar type.

Bibliography:

[1] Semenov E I 1986. *Forging and Punching. Directory* Volume 2 271 p

[2] Grezina I V 2008. Development *of a car half axle blanking technology in a horizontal forging machine with directional fiber structure* (Education & Science, electronic edition)

[3] Bryukhanov A N 1975 Forging and blanking. Textbook for higher educational engineering establishments: 299 p.

[4] Birger I A Mavlyutov R R 1986 *Strength of Materials*: Textbook: 435 p

[5] Somkina A S 2013 *Reasons for blank upsetting with the transcendent value of the upset length ratio by means of of mathematical modeling* (Youth Science and Technology Bulletin, an electronic resource)

[6] Zhelezkov O S, Gatin F F and Malakanov S A 2013 *Upset blank stability increasing while tie rod blanking in a horizontal forging machine* Modeling and development of metal forming processes pp 60-67

[7] Gatin F F 2015 *Improvement of multitransition hot blanking of boss bars in a horizontal forging machine.* (Thesis for the Doctor of Philosophy in Technics, Magnitogorsk): 124 p

[8] Pankratov D L and Valiev A M 2014 Creation of anisotropic friction conditions during the process of hot forging *IOP Conference Series: Materials Science and Engineering* **69** 012034

Plasma-jet hard-facing modeling

V V Zamorskiy

"Automation and Control" Chair, Associate Professor, Naberezhnochelninsky Institute (branch of) Kazan (Volga Region) Federal University

e-mail: altairvid@mail.ru, VVZamorskij@kpfu.ru

Abstract.The problem of automating the process of parts restoration by means of plasma welding in the environment of ANSYS.

1. Introduction

In order to restore and / or strengthen powder plasma surfacing is applied in mechanical engineering. Advantages of this technology are such as high performance, quality of weld metal, minimal loss of filler material, the possibility of facing relatively thin layers. [1, 6]

The problem is its widespread introduction comprises the difficulty of using traditional design methods, because the solution of such problems requires a large number of field experiments. This is due to large material costs and time-consuming processes when preparing the process of parts restoration. A particularly large number of samples is required in assessing the stability of the welding parameters to random deviations and determining tolerances. Implementation tasks for CAE modeling environment will automate the process of parameters optimization on the stage of technological preparation. [3]

2. Mathematical model r construction

The model considers the propagation of energy along the axis x, wherein the beam axis coincides with the x axis of the cylindrical coordinate system x, R (see fig.1). The surface of the metal substrate is normal to the x-axis at $x = x_3$. On the surface x_2 separating the condensed phase and running over a two-phase medium (fig.1) it follows the law of energy flow conservation, taking into account the energy absorption of the radiation surface, transferring heat energy particle surface, the heat transfer surface by convective heat and radiative cooling and heat sink deeper.

The equation of motion is based on the implementation of the law of mass surfaced particles conservation. At the same time account evaporation and condensation of steam are not taken into account at the condensed phase boundary x_2. On the boundary x_3 between weld material and the substrate material the equality of temperature and heat flow continuity are followed. At the boundary x_4 between the solid and molten materials it provides temperature conservation and the law of heat flow conservation based on energy consumption for melting, when $x_4 <> x_3$.

IOP Conf. Series: Materials Science and Engineering **134** (2016) 012045 doi:10.1088/1757-899X/134/1/012045

The system of equations, describing the process of dispersed metal particles plasma-jet hard-facing when moving in a stream of inert gas, on the fixed metal substrate contains the following basic equations:

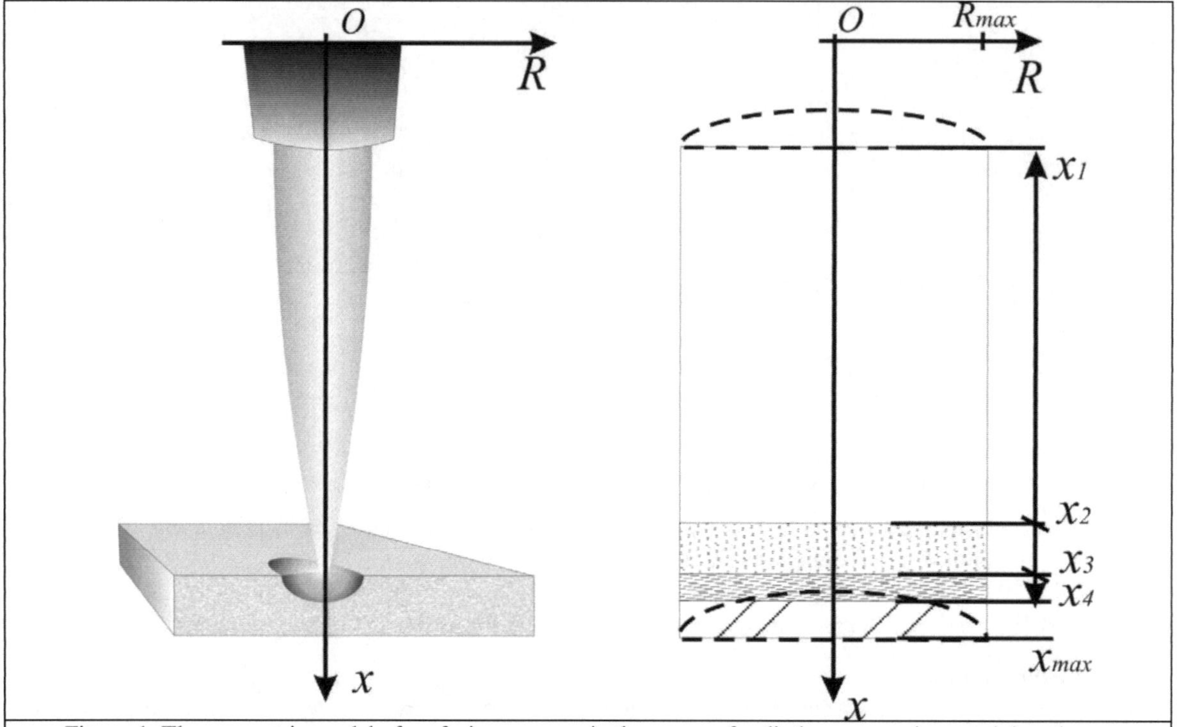

Figure 1. The geometric model of surfacing: $x_1 - x_2$ is the scope of radiation on moving particles, the arrow shows the direction of radiation propagation and particle motion; x_2- x_3 is weld material; x_4 is molten material substrate; $x > x_4$ - solid substrate

Quasi-stationary equation of energy transfer in the single-scattering energy particles approximation $$\frac{\partial q_n}{\partial x} + (\alpha_0 + \alpha_{op} + \alpha_M) = 0,$$ where $q_n = IU_\partial \dot{\eta} / v_\partial$ – heat input, J/cm $U\partial$ - arc voltage, I- amperage, $\dot{\eta}$ -utilization of energy roof for penetration the deposited material; v_∂ - deposition rate, cm/s, $$\alpha_0 = \pi N_0 \int_0^\infty r_0^2 (K_{\ddot{I}} + K_\delta) f_0(r_0) dr_0$$, attenuation coefficient of particle radiation, teplomasso halos and gas phase , when α_{op}, $\alpha_M << \alpha_o$ and they are neglected, K_n, K_p, K_o respectively, the factors of absorption efficiency, scattering and attenuation with the current particle radius r_0, f_0, (r_0) - a function of the particle size distribution, $\int_0^\infty f_0(r_0) dr_0 = 1$,	(1)

The energy balance equation [5] $$\rho_0 V_0 c_0 (\frac{\partial T_0}{\partial t} + (V_x \nabla) T_0 = \frac{1}{4} q_n K_n S_0 - \bar{j}_m S_0,$$ *where ρ_0, c_0 - density and specific heat of the particles material, T0 - temperature of the particles, assumed to be uniform throughout its volume, t - time, $V_0 = 4\pi r_0^3/3$, $\bar{j}_m$, $\bar{j}_e$ - respectively, density, mass and energy flows on the surface of the particle because of its heat and mass transfer from the gas, Vx - velocity of the particle along the x axis,*	(2)

The two-dimensional heat equation of the gaseous medium [4] $$\rho_c c_c (\frac{\partial T_c}{\partial t} + (U\nabla) T_c) = \frac{\partial}{\partial x}(k_c \frac{\partial T_c}{\partial x}) + \frac{1}{R}\frac{\partial}{\partial R}(k_c R \frac{\partial T_c}{\partial R}) + q_T,$$ T_c - ambient temperature, $q_T = 4\pi N_0 \int_0^\infty [\bar{j}_T + \bar{j}_m c_1 (\bar{T} - T_c)] r_0^2 f_0(r_0) dr_0$ - power density of heat in the medium due to heat conduction from the particles of J_T and heating with superheated steam, c_c, c_1, c_2 - accordingly, the heat capacity of the gas-steam and gas, T - temperature of the medium directly from the surface of the particle, U - gas stream velocity, $\bar{j}_\varepsilon = \bar{j}_{рад} + \bar{j}_T + \bar{j}_m (c_1 (\bar{T} - T_c) + L_И)$, L_v – latent heat of vaporization of the material particles, $\bar{j}_{рад} = \varepsilon\sigma(T_0^4 - T_\infty^4)$ - energy flux density of rejection heat radiation from the particle, where ε - the emissivity of the particle surface, σ - Stefan-Boltzmann constant, κ_c - thermal conductivity of the medium,	(3)

The equation of state of the gas medium - $$p_c = R_r \rho_c T_c = p_\infty$$ $\rho_c = \rho_1 + \rho_2$, ρ_1, ρ_2 - in accordance with it, the density of the gas-vapor medium, particles gas and inert gas, p_c — pressure, $R_Г$ - gas constant, r_∞, $\rho_{2\infty}$, f_∞, T_∞, ρ_∞ - the initial values of the relevant parameters	(4)

The heat equation for the condensed phase of the deposited material and the substrate material $$\rho_П c_П \frac{\partial T_П}{\partial t} = \frac{\partial}{\partial x}(k_П \frac{\partial T_П}{\partial x}) + \frac{1}{R}\frac{\partial}{\partial R}(k_П R \frac{\partial T_П}{\partial R}),$$ ρ, $c_П$, $\kappa_П$ — respectively, density, specific heat and thermal conductivity of the condensed phase at $x \geq x2$ $(x3)$, taking into account the differences of these values to the deposited material and the raw material substrate - $T_П$ - temperature condensed phase	(5)

with the initial conditions

$T_0(x, R, t=0) = T_\infty$,	$T_c(x, R, t=0) = T_\infty$,	$T_Ï\ (x, R, t=0) = T_\infty$,	(6)
$r_0(x, R, t=0) = r_\infty$,	$\rho_1(x, R, t=0) = 0$,	$\rho_2(x, R, t=0) = \rho_{2\infty}$,	
$f_0(x, R, t=0, r_0) = f_\infty(r_\infty)$			
где r_∞, $\rho_{2\infty}$, f_∞, T_∞, ρ_∞ - the initial values of the relevant parameters			

and boundary conditions

$T_0(x = x_1, R, t) = T_c(x = x_1, R, t = 0) = T_Ï\ (x = x\max, R, t) = T_\infty$	(7)

$k_c \frac{\partial T_c}{\partial R}\|_{R=0}=0$, $k_{\ddot{I}} \frac{\partial T_{\ddot{I}}}{\partial R}\|_{R=0}=0$,	
$\rho_c D_c \frac{\partial(\rho_1/\rho_2)}{\partial R}\|_{R=0}=0$, $q_n(x=x_1, R \le R_{\max}, t)=0$,	
$T(x<x_2, R=R_{\max}, t)=T_{\ddot{I}}\ (x=x_{\max}, R, t)=t_{\infty}$, where radius R_{max} and length x_{max} for the calculated volume are selected so that unperturbed initial parameter values remain at the boundaries in the time interval under consideration, $\rho c = \rho 1 + \rho 2$, $\rho 1$, $\rho 2$ - respectively, gas-steam density, particulate material steam and inert gas, k_c, Dc – thermal conductivity of the medium, and medium vapor diffusion, Tc - ambient temperature	
$\rho_1(x, R=R_{\max}, t)=0$,	
$[A_{\Pi} I + N_0\rho_0 V_x \int_0^{\infty} [c_0(T_0 - T_{\Pi}) + L_{n\pi}] \frac{4}{3}\pi r_0^3 f_0(r_0) dr_0 -$ $\alpha_T(T_c - T_{\Pi}) - \varepsilon\sigma(T_{\Pi}^4 - T_{\infty}^4)]\|_{x_2-0} = -k_{\Pi} \frac{\partial T_{\Pi}}{\partial x}\|_{x_2+0'}$ $-k_{\Pi} \frac{\partial T_{\Pi}}{\partial x}\|_{x_3-0} = -k_{\Pi} \frac{\partial T_{\Pi}}{\partial x}\|_{x_3+0}, -k_{\Pi} \frac{\partial T_{\Pi}}{\partial x}\|_{x_4-0} = -k_{\Pi} \frac{\partial T_{\Pi}}{\partial x}\|_{x_4+0} -$ $L_{\Pi}\rho_{\Pi} \frac{dx_4}{dt}$, where An - absorbance of the condensed phase surface; Vx - velocity of the particle along the x axis, ρ_0, c_0 - density and specific heat of the material particles, Tn - condensed phase temperature, $L_{n\pi}$ - melting heat, α_T - the coefficient of convective heat transfer, ρ, c_n, κ_n - respectively, density, specific heat and thermal conductivity of the condensed phase at $x \ge x2$ $(x3)$, taking into account the differences of these values differences for deposited material and initial support material	
$T_{\ddot{I}}\|_{x_3-0} = T_{\ddot{I}}\|_{x_3+0}, T_{\ddot{I}}\|_{x_4-0} = T_{\ddot{I}}\|_{x_4+0} = T_{\ddot{i}\ddot{e}}$,	
$0 \le t < t_н : x_2 + x_3,\ t \ge t_н : x_2(t) = x_3 - \frac{w_0 V_x (t - t_н)}{\rho_0}$, where $w_0 = \frac{4}{3}\pi N_0 \rho_0 \int_0^{\infty} r_0^3 f_0(r_0) dr_0$ - particulate material bulk, contained in a medium volume unit, t_H - time of beginning deposition of powder initial time deposition onto the substrate	

The melt surface shape (r - meniscus radius) can be determined from the equations of equilibrium pressures:

$p_\sigma + p_{час} + p_g + p_{вн} = 0$	(8)

p_σ - capillary pressure, which depends on the melt surface curvature $p_\sigma = (2\sigma)/r$, σ - interfacial tension, $p_{час}$ - the pressure of the powder particles, depending on the speed and flow, p_g - hydrostatic pressure $p_{вн}$ - internal pressure, which is determined by appropriate amounts of deposited metal and powder entering the molten bath.

2.1. Border conditions

IOP Conf. Series: Materials Science and Engineering **134** (2016) 012045 doi:10.1088/1757-899X/134/1/012045

Plasma energy values q_0 are approved for heating and melting of the powder particles and the surface layer of the substrate, but without evaporation. As metal evaporation intensively occurs at temperature $T \approx\geq T_{КИП}$, where $T_{КИП}$ is a metal boiling point at atmospheric pressure, then at a considerable range of temperatures $Tпл \approx\leq T < Tкип$ modes without evaporating particles and the substrate can always be realized. In this case $\bar{I}_\epsilon$, qT the mass flow density $\bar{I}_\epsilon = 0$ that means particles and the surface of the substrate only exchange heat with the environment.

The motion of particles in the gas stream is not considered as the particle velocity Vx is taken to be the gas velocity U.

Equations (1), (2), (3) and (5) can be regarded as one-dimensional along the coordinate x as values gradients along x are much greater than the corresponding gradients along R and growth examination of the deposited layer along the x flow axis is of the greatest practical interest.

3. Modeling tasks decision

С точки зрения проектирования технологического процесса плазменной наплавки обычно решаются следующие физико-математические задачи: In terms of plasma deposition design process the following physics-mathematical problems are usually resolved:

- the ratio plasma energy parameters, gas flow and sprayed powder rate, deposition rate
- plasma distribution in gaseous atmosphere (the ratio of temperature and particle size)
- the motion trajectory of the powder particles formation
- plasma thermal effect on a work piece (warming depth)
- powder and parts interaction (mixing)
- surface shape of the melt formation

Thus, model (1) - (8) describes the phenomena discussed in various physical disciplines. This shows that the development and debugging of software for design automation is a hard-working and long-length task. Development of computer design methods has led to the creation of interconnected systems CAD for designing technologies and tests.

Currently, there are a large number of Russian and foreign aided CADs (Computer-Aided Design) systems. They are usually used in conjunction with engineering calculations automation and analysis systems (CAE), which allow to carry out the tasks of a process modeling in different combined environments. The most common CAE-systems include ANSYS, LS-DYNA, Nastran, Abaqus, T-FLEX CAE, Deform, Qform and so on. It should be noted that CAE systems are divided into highly specialized (e.g. Deform and Qform) and a wide range of applications motivated, such as ANSYS (ANSYS, Inc.), Nastran (MSC.Software Corporation), Abaqus (Abaqus, Inc.). The most popular software packages, focused on a wide range of tasks are products of MSC.Software Corporation (Marc, Nastran, Patran), ANSYS, Inc. (ANSYS, LS-DYNA, and so on).

A distinctive feature of ANSYS software is the ability to tale mono and multidisciplinary calculations. In addition, the problem range of this package is much wider than with its counterparts. Therefore, despite the relatively high cost, the preference for the objectives of the study of plasma-powder surfacing has been given to ANSYS.

However, ANSYS, as well as the above-listed CAE system cannot satisfactorily issue and calculate the parameters of welding process with the given accuracy. Settlements in CAE systems are manufactured by highly simplified patterns and the needs of engineers are sometimes neglected because of their diversity. This has led CAE developers to create sophisticated subsystems setting calculation formulas and software environment for interdisciplinary association of calculations.

Therefore, to adjust the analyzing system you need to create an algorithm for determining the optimal conditions for quality powder surfacing parts of engineering production. Then, develop a methodology to assess the stability of the surfacing, the probability of obtaining the desired quality, and to develop a method for determining the tolerances for the process parameters.

A feature of the solution is a representation of a physical space as consisting of two phases, with the free surface - metal and gas. This makes it possible to take into account the effects of surface tension, to simulate the adhesion. It is based on the solution of conventional and radiation heat transfer problems. Multicomponent metal phases are considered (solid-liquid) as well as gas flows effects (argon gas and metal vapor) and diffusion and melting are emitted.

Electric arc electromagnetism problem solution allows to control the temperature, taking into account gas properties in the local thermodynamic equilibrium, and to describe the thermal and mechanical collision [2].

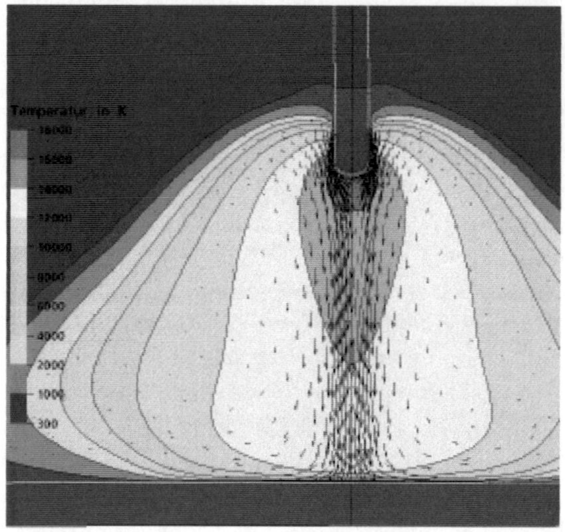

Figure 2. The result of plasma jet with deposited particles simulation

The joint solution of electromagnetism and melting problems accounting surface tension makes it possible to obtain the geometry of a weld pool, to analyze structural changes in a detail surface and molten droplets spraying.

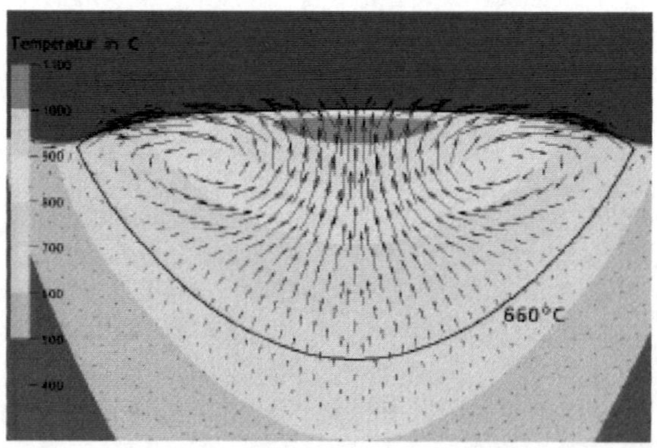

Figure 3. Model Weld pool model

Flow rate modelling allows to show gas flow rate in full detail, as well as the heat transfer, diffusion, mixing the shielding gas and molten filler powder, to take into account the effect of the electric arc on the transport of the material point and the spread of smoke.

IOP Conf. Series: Materials Science and Engineering **134** (2016) 012045 doi:10.1088/1757-899X/134/1/012045

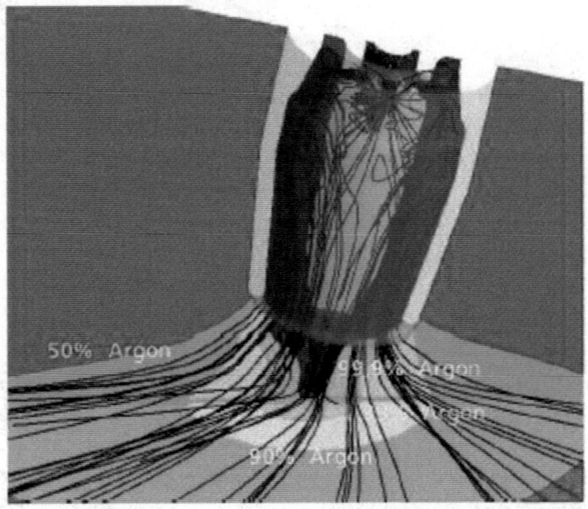

Figure 3. Powder flow in the plasma modelling

References

[1] Knyazeva A G 2009 *Modeling Physical and chemical phenomena modeling in the processes of materials surface treatment with the help of high-energy sources.* (Mathematical modeling of systems and processes. -. № **1**) pp 66-84

[2] Spille-Kohoff A 2009 *Electric Arc Simulation with Electro-Magnetics beta-feature of ANSYS CFX* (Germany: ANSYS Conference & 27th CADFEM Users' Meeting - November 18-20, Congress Center Leipzig, Germany)

[3] Kuryaninova E I and Yerofeyev V A 2007 *Optimizing of laser gas-powder surfacing technological parameters based on mathematical modeling* (Welding Production, №4) p. 21-25

[4] Grigoryants A G , Shiganov I N and Chirkov AM 2004 *Hybrid technologies* (Moscow: Publishing House of MSTU named after Bauman)

[5] Tikhonov A N, Samarsky A A 1999 *Mathematical physics equations* (Moscow: Studybook. The sixth edition: Moscow State University Publishing House)

[6] Gavrilova V A, Kashapov N F and Kashapov R N 2011 *Plasma application of protective polymer-powder coatings to ultrasonic sensors* Biomedical Engineering **45** 198–200

[7] Israfilov I H, Galiakbarov A T, Bashmakov D A, Gabdrahmanov AT and Samigullin A D 2014 Pulse plasma surface thermostrengthening of machine parts *IOP Conference Series: Materials Science and Engineering* Volume **69 012037**

IOP Conf. Series: Materials Science and Engineering **134** (2016) 012046 doi:10.1088/1757-899X/134/1/012046

Modification of the surface of parts

L A Simonova, M A Chernova and V V Zvezdin

Department of Energy and Information, Kazan Federal University, Naberezhnye Chelny, Russian Federation

smilby@mail.ru

Abstract. The increasing wear resistance of parts is one of the main tasks of engineering. The achievement of stable parameters is a difficult task. It depends on capabilities of equipment, type of evaporator, interaction evaporator with the evaporating substance and deposition process.The control system must foresee and identify emerging cracks or damage of the system components. At every stage software of control process system must automatically make the optimal decisions based on information about the process. It also improves the efficiency of the operator. This article focuses on the results of using intelligent control system technological process.

Key words: intelligent control system, modification the surface of the parts, wear resistance.

Introduction:

Management of technological complexes coating caused by the need to obtain the desired quality parameters of the process and provide the required physical and chemical properties. Modification of the surface layer of parts and tools is widespread in engineering [1]. This allows giving the surface the necessary quality by specified performance requirements [2]. Introduction of automated control systems of technological complexes of surface modification leads to high rates of quality and stability of the parameters of the surface layer parts. This is justified by the high performance and efficiency of the coating process.

Using previous experience on the analysis of deviations from the nominal values of process parameters is possible using elements of artificial intelligence [3].

Materials and methods:

The effectiveness of control systems of technological complexes of surface modification to a large extent determined by the scientific validity of the decisions and timely control actions on the actuators. The solution to this problem allows the use of the mechanisms of artificial intelligence [4]. At every stage software of control process system must allow automatically make the optimal decisions based on information about the process.

The wear resistance is the result of a complex set of processes that occur not only in the physical wear of the tool edges, it may be accompanied by plastic deformation, fatigue, and even the diffusion processes. The wear resistance depends not only on the structure and properties of the tool steel, but also on the properties of the processed material (its hardness), and the coefficient of friction and the external environment in which the processing occurs. Increasing the hardness tends to reduce viscosity. If dynamic load is dominated in the tool, for the increase of wear resistance for a given hardness and content of carbides the large supply of viscosity is required . In many cases, increasing the viscosity should seek to reduce the hardness, increased surface hardness improves the wear resistance.

To achieve the required quality of the technological process (the depth of the modified zone tool wear, surface roughness, micro-hardness, chemical composition of the deposited layer zone, etc.), depending on the parameters of the technological complex, it is necessary to use control system.

Results:

The results of experimental studies of laser hardening, welding carbide cutting edge of the tooth powders hob are shown. The AL2O3 sprayed onto substrates of different materials for experimental research. The installation UVN-4M and Eifeler alpha 400 p are used As a technological complex. Proposed intelligent control system and methods for its implementation have been applied in vacuum sputtering equipment UVN-4M. It is intended to magnetron metals sputtering on the surface of products of various geometric shapes, open to a direct hit on it evaporated metal vapor. The principle of the device is based on the sputtering material by ions of the working gas. Is formed in an abnormal glow discharge plasma in crossed electric and magnetic fields in working chamber.

Figure 1 (a, b, c,) are photographs of sputtered layers on glass.

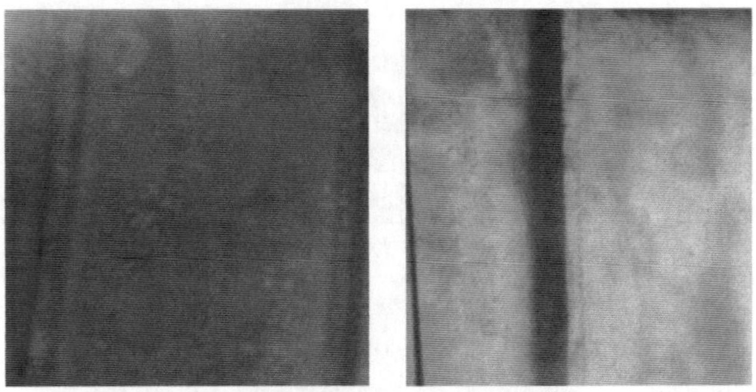

a) b)

Figure. 1 - before the application of the system: the sample of a - layer thickness is less than specified, the sample b - thickness of the layer is greater than a given;

a)

Figure. 2 after application of the control system: the sample a - thickness of the sprayed layer is within the required limits.

The level of adhesion on test specimens by direct separation on the device Elcometer 108 according to ISO 14923: 2003 and ISO 2063: 2005 was measured. On polished and degreased surface of the coating with an adhesive M2000 were glued samples, then the hydraulics system of device apply a force whose value was fixed at the time of separation. The best adhesion was observed in the sample where used the control system.

Experimental studies on the surface modification carried out on the laser technological complex pulse-periodic action LRS-150 A. Pulse duration: t = 3 ms. Energy pulse: E1 = 11 J.

Microhardness the structure of parts was measured by GOST R ISO 6507-1-2007 on Hardness MicroMet 5104 under a load of 1 H. For the material of parts is used tool steel R18K5F2.

Modifying the surface of a layer of steel - an alloy powder PG - CP4 - OM. It is mainly used to cover the parts of machines, which are subjected to abrasion with a relatively low impact load.

The microstructure of the base metal plate is a ferrite and structural - free cementite 0V points GOST 5640-68 with microhardness up to 390 HV0,1. The microstructure of the hardened surface layer with carbides. High hardness layer has a positive effect on the wear resistance of the tool. Hard metals with high hardness, exhibit low ductility and are fragile, therefore have poor impact resistance. Friability hardmetals stored at high temperatures, so that they are prone to cracking during surfacing . Figure 3 shows a fragment hob with laser welding.

Figure 3. - Fragment hob with laser welding.

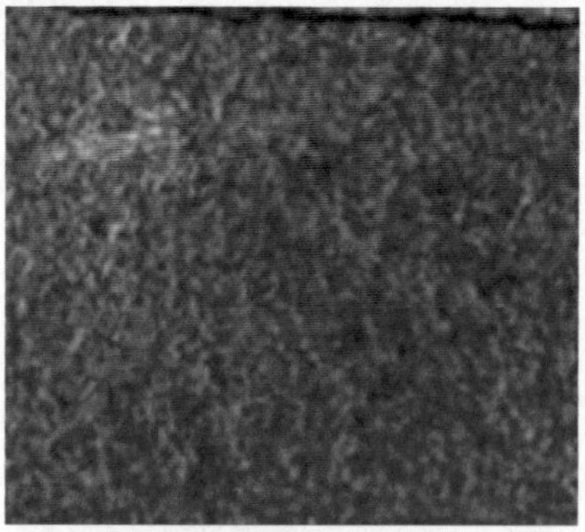

Figure 4 is a sample of the powder-coated alloy PG - CP4 - OM.x 5.

Conclusion:

From a comparison of different methods of changing the microstructure of the metal, the use of laser deposition welding and using alloy powders optimally to create a modified surface of the part.

The main influence on the process of laser welding and plasma spraying of coatings with high resistance to wear have enhanced physicochemical properties of both the materials and the environment in which the process occurs. This difference is due to the distribution of thermal field in a variety of materials characterized by different chemical activity and thermal properties.

Analysis of the relationship of parameters of technological systems and quality of deposition and deposition showed that a great influence on the quality of their stability and optimal values and all parameters are temperature dependent and physico-chemical properties of the medium.

As a result of implementing the system was to improve the quality of decisions on the management of the process and reducing the time for troubleshooting. This greatly improves the efficiency of the coating process by reducing the time for diagnosing faults using control system with elements of artificial intelligence.

Designed technological complex control method using fuzzy logic to maintain a predetermined process parameters coating enhanced the abrasion resistance components to 20%.

REFERENCES:

[1] R M Khisamutdinov, M R Khisamutdinov 2014 Automation system goals for the creation and operation of the tool *IOP Conference Series: Materials Science and Engineering* **69** 012021

[2] S M Portnov, R R Saubanov, R A Kish, I N Kuznetsov, I H Israfilov, V V Zvezdin, A I Nugumanova 2011 The process control system of heat treatment with concentrated flows of energy surfaces of parts *Global scientific potential - scientific journal* 8 pp 95-100

[3] L A Simonova, E I Egorova 2015 Development of structural element precedent of technological process in computer-aided *design IOP Conference Series: Materials Science and Engineering* **86** 012017

[4] K V Klochkova, S V Petrovich, L A Simonova , L R Yusupov 2015 Development of methodology for controlling the parameters of *TP IOP Conference Series: Materials Science and Engineering* **86** 012015

[5] Chernova M A, Simonova L A, Israfilov D I, Nugumanova A I 2012 *Developing the knowledge base for the management of the technological complex vacuum sputtering installation UVN-4M* European applied sciences. ORT Publishing vol 1 pp 331-334

[6] Fedorov Y N 2008 *Engineer directory for process control: design and development* Moscow: Infra-Engineering pp 928

[7] Tarasov V B 2002. *From multi-agent systems to intelligent organizations* M: Editorial URSS pp 352

[8] Makarov I M, V M Lokhin, S V Manko and M P Romanov 2006 *Intelligence and Intelligent Control Systems* Publisher: Nauka pp 333

[9] Saaty T L 2008 *Decision-making at the accuracy, improves raw material stocks. dependencies and feedbacks*. Analytic Network. M

[10] KashapovL N, KashapovN F and Kashapov R N 2013 *Investigation of the influence of plasma-electrolytic processing on the surface of austenitic chromium-nickel steels.* IOP Conference Series: Materials Science and Engineering *J. Phys.: Conf. Ser.***479** 012003

IOP Conf. Series: Materials Science and Engineering **134** (2016) 012047 doi:10.1088/1757-899X/134/1/012047

Research of influence of geometrical parameters of a stamp engraving of setting transitions at setting of semi finished pipe products to receiving defect-free products

D L Pankratov[1], R S Nizamov[2] and I Zh Harisov[2]

[1] Kazan Federal University, Naberezhnye Chelny Institute
Pr. Mira 68/19, Naberezhnye Chelny, Republic of Tatarstan, Russia, 423812 ,
pankratovdl@gmail.com
[2]Forging Plant, Public Joint Stock Company «KamAZ», Naberezhnye Chelny,
Avtozavodsky Av. 2, 423827, Republic of Tatarstan, Russia

e-mail: nizamovrs@yandex.ru

Abstract. The cause of a pipe clip during the setting of the pipe is determined on the basis of modeling. The method, which excludes the formation of a defect in a semi finished product by changing the configuration of the setting transition, is developed.

Forgings of pipes for setting (group V) have the following most common options for setting transitions (wall thickening t) [2] :

- by increasing the outer diameter d (Figure 1, a);
- by reducing the inner diameter d0 (Figure 1, b);
- by changing both the diameters d and d0 (Figure 1, c).

IOP Conf. Series: Materials Science and Engineering **134** (2016) 012047 doi:10.1088/1757-899X/134/1/012047

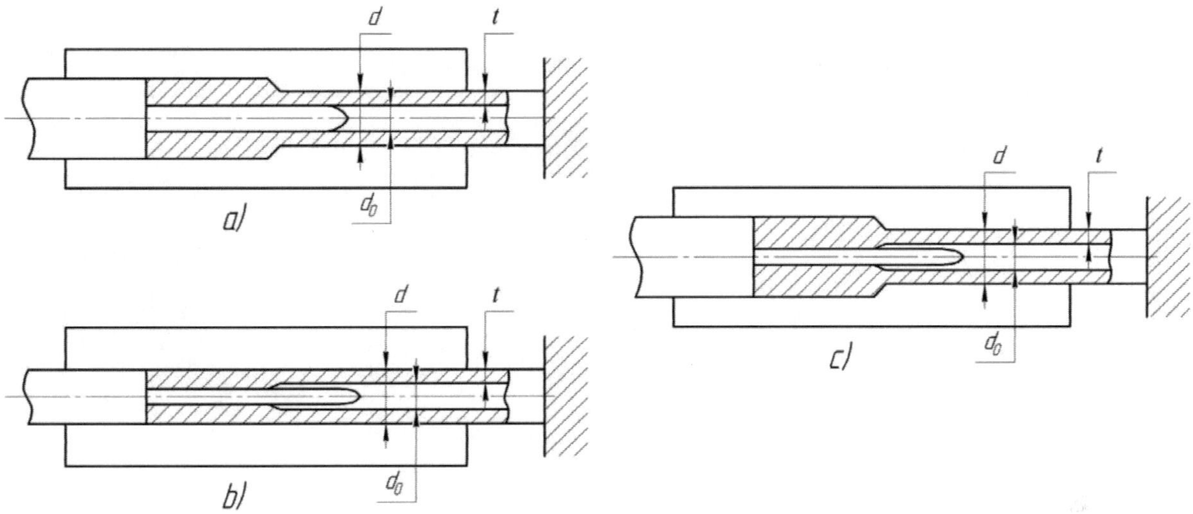

Figure 1. Options of pipe setting.

Practicability of a pipe setting method depends on specific conditions: the work piece material, the available equipment, given technical demands to stampings concerning the form, the size of the semi finished pipe, etc.
Technological process of manufacturing semi axle housing in a Forge plant PJSC KAMAZ is a setting of 3 transitions from a pipe Ø121 × 22 mm of 35HGSA brand. Calculation of operating technology transitions are based on the recommendations contained in the directory "Forging and Stamping" [2]. The main problem of this process, periodically occurring when setting, is a defect in the form of cracks in the inner cavity of the flange portion of the forging. The defect is a clip on the inner surface of the forging at the middle portion of the set pipe (Table 1). The depth of the crease formation is 35 - 40 mm and is more than the actual allowance for further machining.
Location of the clip is difficult for visual identification, which increases the risk of getting the details of semi axle housing with a deviation into the assembly of a finished vehicle and its damage during its exploitation.
The actual depth of the clip can be determined only by a method of a test cleaning. Thus, using the current technology of setting, 100 % of the products should be subjected to an additional operation - cleaning of the interior surface contour and grading according to the actual depth of the defect.

Research of the problem in practice and modeling the process in QForm2DV 8.0.4 (Table 1) showed that the formation of a clip in a final transition is caused by an undercut in the first transition encircling the inner cavity of a semi-finished product. In the 2nd and 3rd transitions the undercut is stamped as a crease and forms a ring clip.

According to A.N. Bryuhanov's [3] recommendations, at setting due to an increase of the outer diameter wall of a semi finished pipe, it's sufficiently resistant against buckling only if the length of the set portion exceeds the wall thickness (t) of the semi finished pipe (Figure 1) no more than 2,5-3 times. It is extremely difficult to comply with this ratio in practice, because inner and outer diameters are limited by the size of the products, and the set has a significant amount of a material. In case of forging with a representative the height of a set portion is 116 mm, which is more than 5 times larger than the wall thickness of the initial pipe.

IOP Conf. Series: Materials Science and Engineering **134** (2016) 012047 doi:10.1088/1757-899X/134/1/012047

Table 1. Comparison of theoretical and practical results.

	Theoretical transition	Modeling in QForm2D	The cut of a semi-finished product
Transition 1	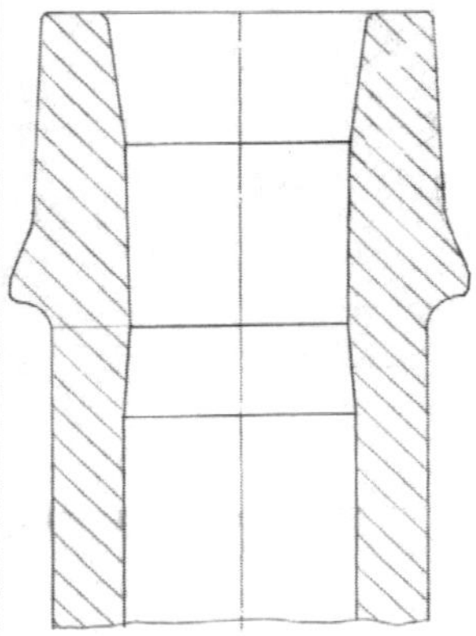	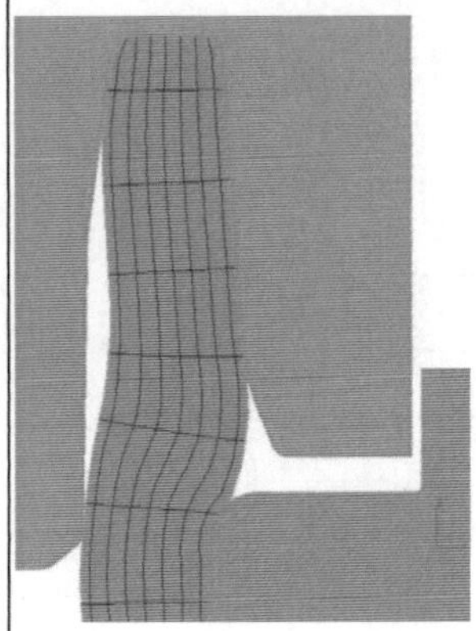	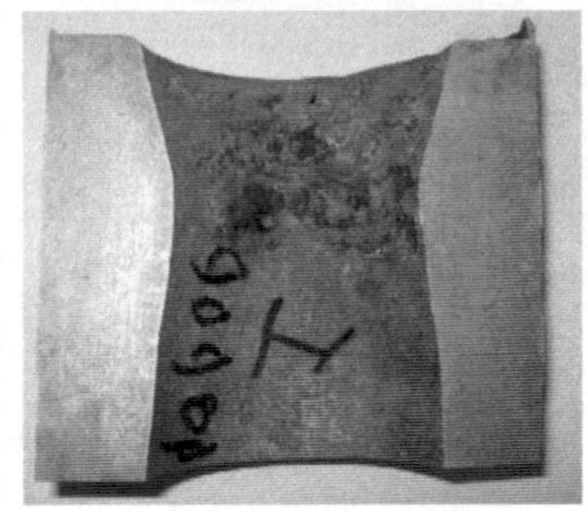
Transition 2	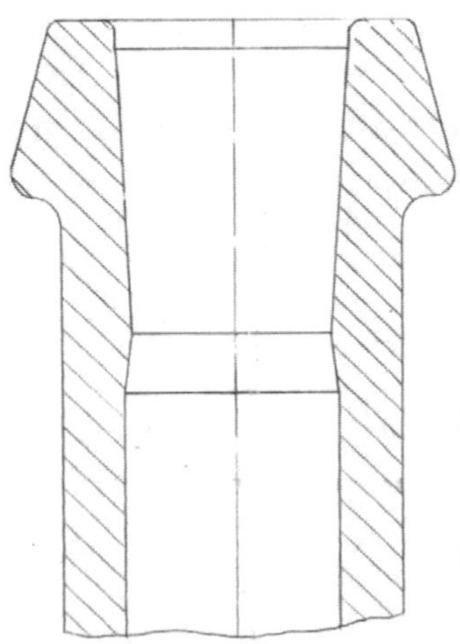	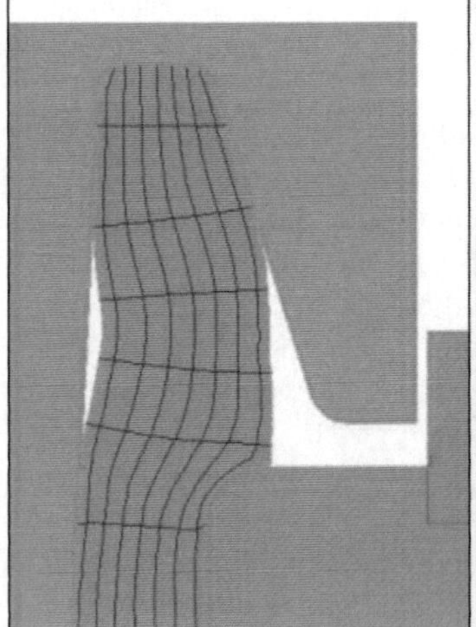	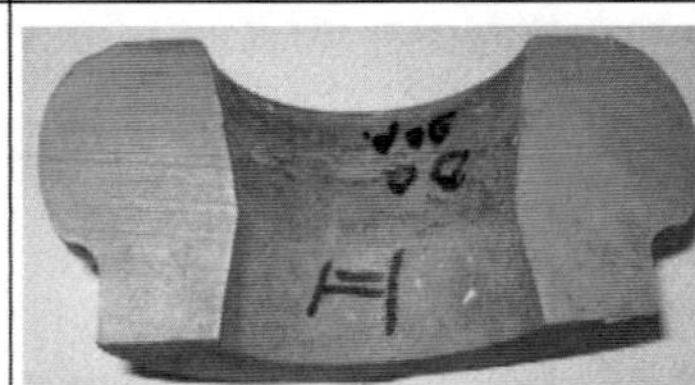 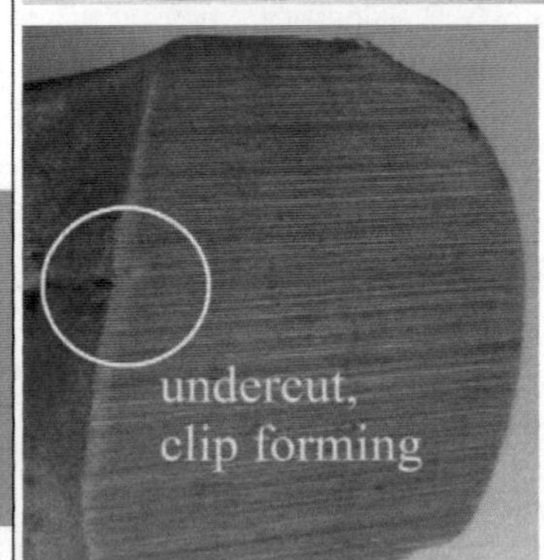

Transition 3	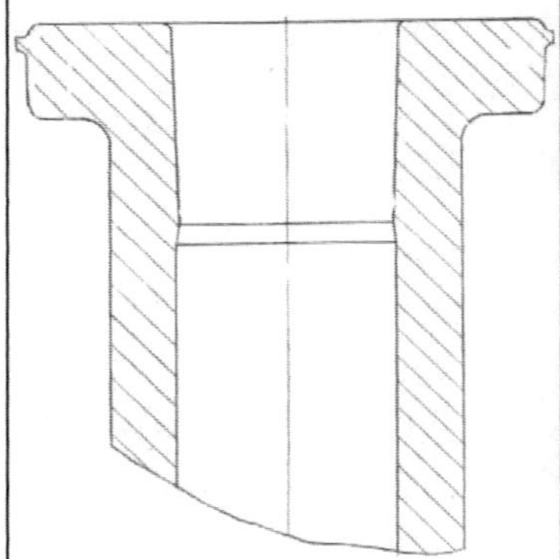	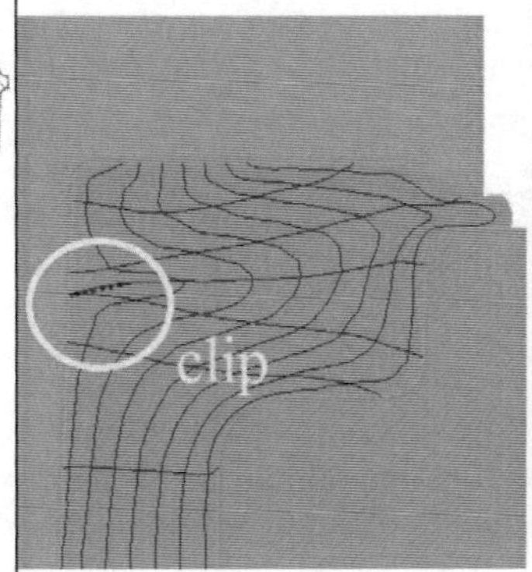	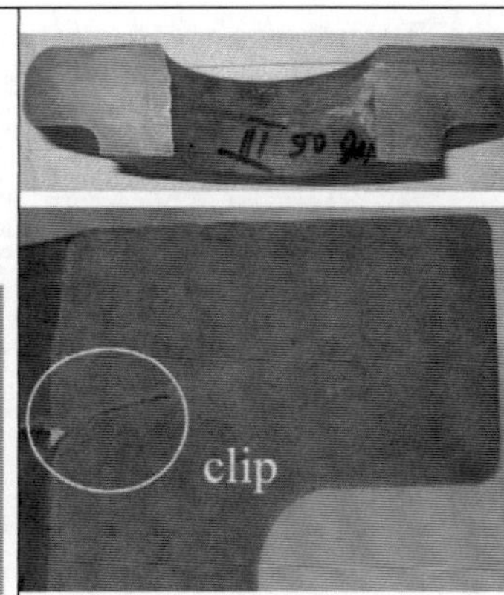

To analyze the process of a semi finished pipe forming, the process of modeling of free 30% settling of the initial height of the pipe by tools of different configuration (Figure 2) in the software QForm3D V 8.0.4 with the following conditions took place: the parameters are close to real conditions of the task (semi finished steel work piece 37CrMo5, DT = 121 mm, t = 22 mm, the heating temperature of the work piece before stamping T = 1220 ° C, the grease on the engraving of the stamp - graphite + water), DT = lim.

Modeling is carried out for tools with parameters Δ = 0.25 ; 0.5 ; 0.75, base angle of α = 0 °; 15 °; 35°; 45 ° and a punch with radius of curvature R equal to the wall thickness t (Figure 2).

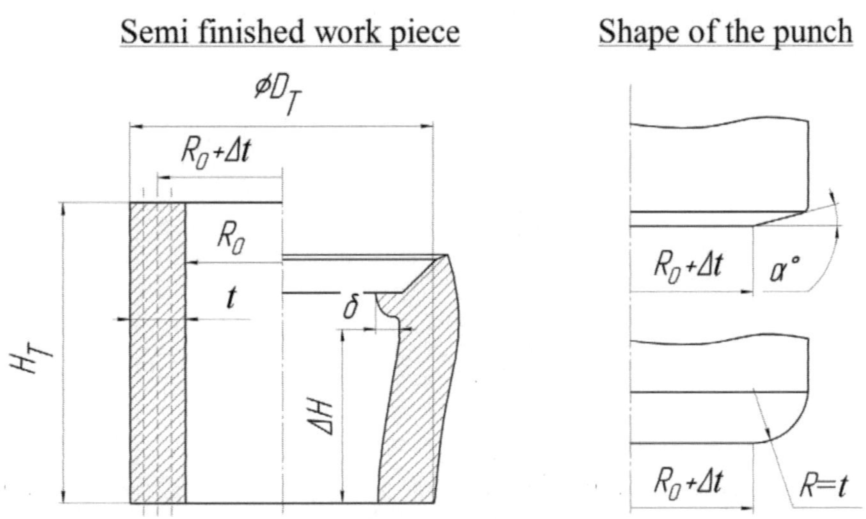

Figure 2. Options of a pipe settling.

It is assumed that the shaping with the tool shown in Figure 2 will minimize the formation of undercuts on the inner contour of the pipe. According to the modeling results (Figure 3) in the case of settling with the tool with parameters R0 + 0,5 t and α = 15 ° the depth of the undercut formation may be reduced by 20% compared with the settling of a punch with a flat butt end α = 0 °.

To prevent the formation of undercuts when setting the 1st setting transition it's proposed to consider the limitation of the semi finished pipe by the outer contour and reduce the diameter of the punch mandrel to provide setting by reducing the inner diameter (Figure 4) [4].

In the current option (Figure 4, a), the mandrel 1 is securely fixed to the insert 2 and a "lock" technology of fixing the upper 2nd and lower 3rd inserts is used, in order to avoid displacement of the upper and lower halves of the stamp setting. In the proposed technology (Figure 4, b) shaping takes place in two stages: the closing of the upper 2nd and lower 3rd inserts, direct setting by the punch 1 moving in the insert-mandrel 2 [5].

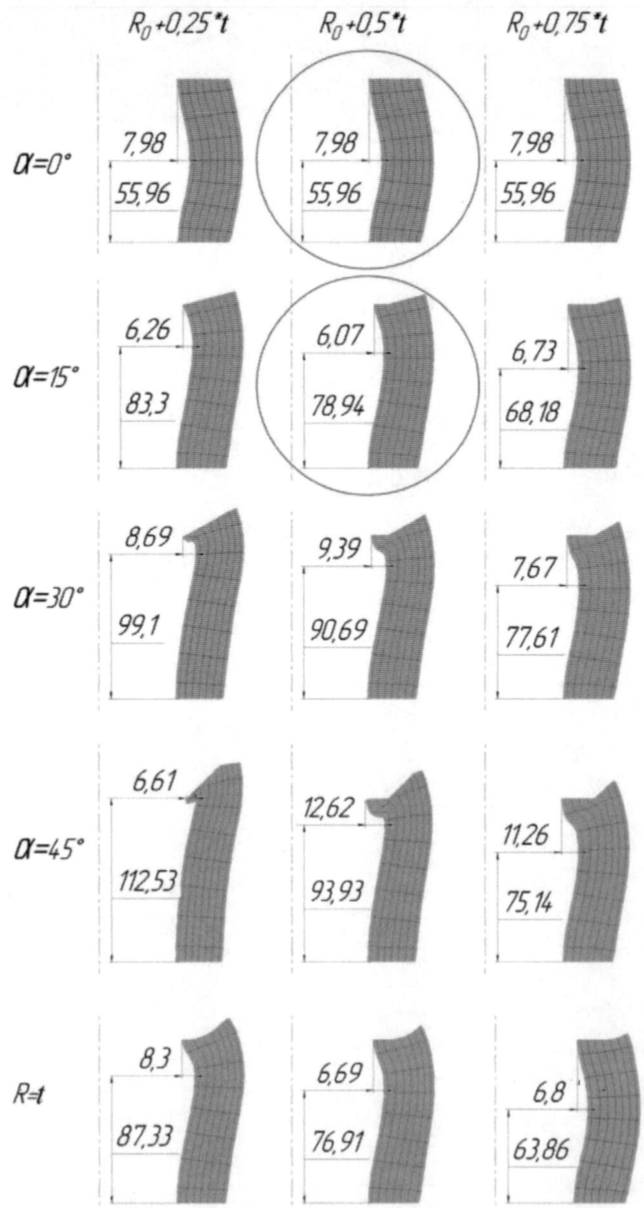

Figure 3. The results of settling modeling with punches of different configurations.

IOP Conf. Series: Materials Science and Engineering **134** (2016) 012047 doi:10.1088/1757-899X/134/1/012047

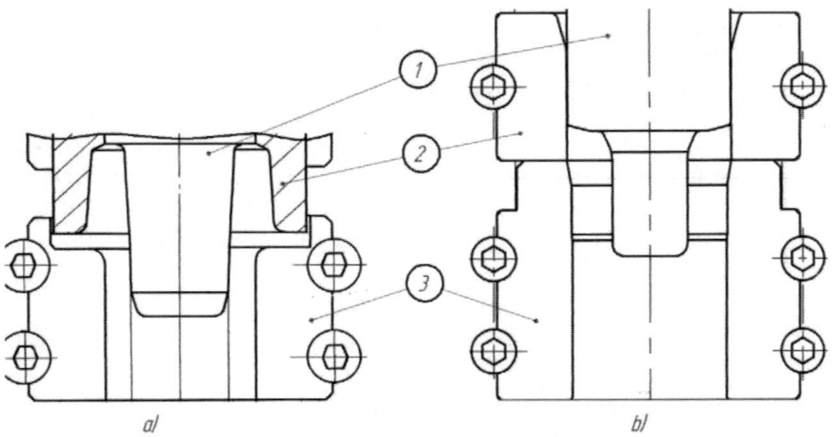

Figure 4. The current and proposed transitions of the1st setting transition.

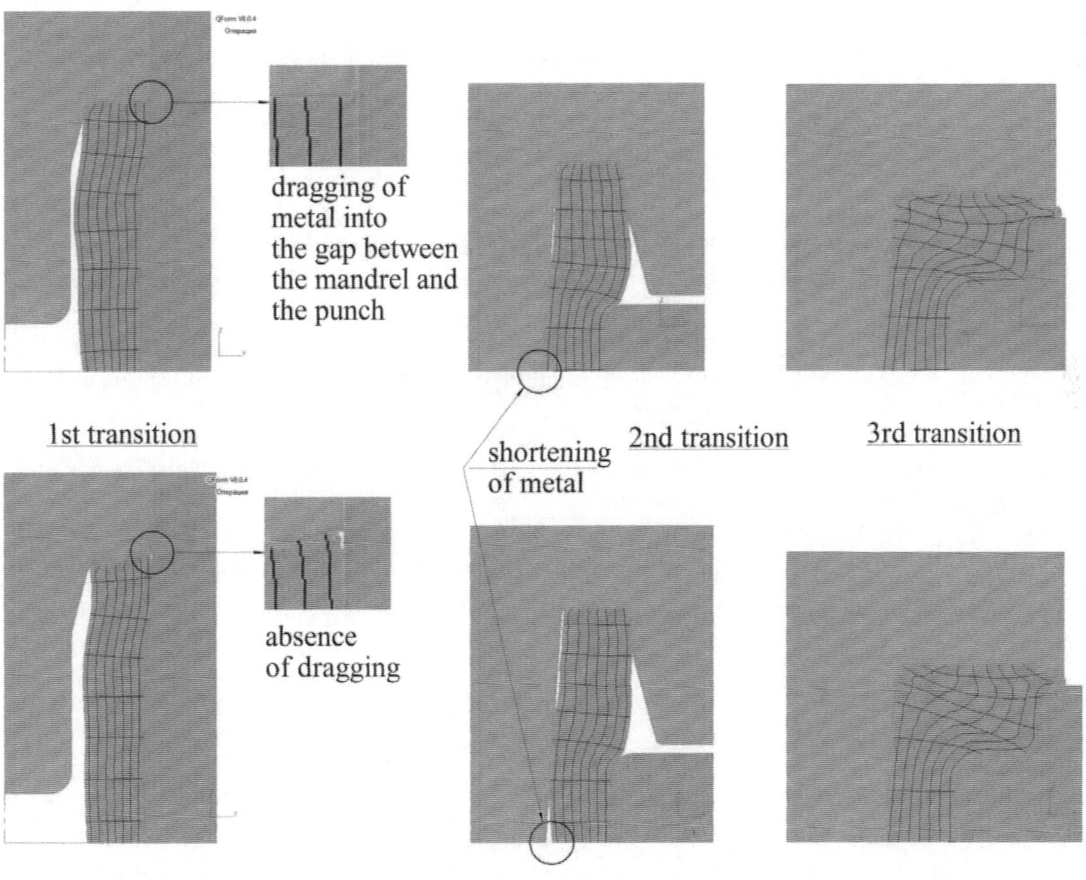

Figure 5. Modeling options for setting of a forging-representative for α = 0 ° and 15 °.

Based on the results of the modeling (Figure 3) further research should be carried out for the second in results tool with a slope of 15 ° (Figure 5). The second and final transition for both versions of setting is the same.

In the first transition the main negative difference between the variant of setting with $\alpha = 0°$ is a significant risk of dragging a burr into the gap between the parts of a die rigging. The die rigging and mutual friction of its elements leads to permanent wear and increase of gaps.

In further transitions this deviation would lead to a defect of a burr stamping into the body of a forging. Studying of the 2nd setting transition indicates the predominance of a settling process when setting with a punch slope of 15 °, which is characterized by less need of stamping force. When setting with $\alpha = 0°$ the excess in the central part of the 1-st transition of metal in the axial direction of the central punch is reduced.

In the 3rd final setting transition a relatively more uniform arrangement of fibers during the setting of the transition with $\alpha = 15°$ is noted.

Based on the above said, for improving the technology of a forging-representative setting of a semi axle housing it is advisable to use a cylindrical punch, having geometrical sizes $\alpha = 15°$ C and R0 + 0,5 t. If necessary, according to the recommendations presented in the paper [6], it's possible to determine the change in power during the setting and to estimate the possibility of cracking.

Conclusion.

The cause of a "clip" defect in the existing technology is analyzed and confirmed by modeling. A method for forming setting transition, aimed at the elimination of a "clip" defect in the finished product in the example of a forging -representative of semi axle housing. The proposed method is applicable to the process of setting of similar forgings from semi finished pipes.

Bibliography:

[1] Storozhev M V 1969 *Technology directory for forging and stamping* pp 523
[2] Semyenov E I 1986 *Forging and Stamping. Directory*. In 4 volumes . - Volume 2. Hot stamping, pp 295
[3] Bryuhanov A N 1975 *Forging and forming*. Textbook for high schools of engineering pp 284
[4] Pankratov D L and Valiev A M 2014 Creation of anisotropic friction conditions during the process of hot forging *IOP Conference Series: Materials Science and Engineering* **69** 012034
[5] Pankratov D L, Shibakov V G, Valiev A M and Valieva RF 2015 Technology of manufacturing of multiple core stamp tools by plastic deformation *IOP Conference Series: Materials Science and Engineering* **86** 012004
[6] Dmitriev A M and Vorontsov A L 2005 *General theory of settling and setting of cylindrical blanks* . Part III. The setting of a hollow blank (Herald of MSTU N E Bauman. The Cycle "Engineering") p125

AR-601 anthropomorphic robot modeling and virtualization toolset for research and education purposes

R Sirazetdinov, A Kamalov, D Nikitina, E Katsevman

Kazan Federal University, 420008, Kazan, Kremlevskaya st, 18

Abstract: A program toolset for modeling and visualization of anthropomorphic robot AR-601 produced by "NPO Androidnaya technika" was implemented, allowing one to test the components of the control system and to debug control algorithms on a virtual model of the robot. The toolset might be used for both scientific and educational purposes. The work is performed according to the Russian Government Program of Competitive Growth of the Kazan Federal University (KFU).

To promote new approaches in engineering education, KFU robots were made remotely accessible to KFU students. This let us to enliven and diversify the education process, deepening their experience in robotic systems[1]

At KFU a software toolset for was developed for educational and scientifical purposes, allowing students to remotely test and debug their works at AR-601 robot (See figure 1)

Fig. 1. AR-601 Anthropomorphic robot

AR-601 produced by Russian company NPO "Androidnaya technika" is a full-sized anthropomorphic robot, a humanoid highly-adaptive platform. The robot is capable to fulfill tasks independently or together with a human, and is able to carry on simple controlled dialogue, to move and orient in space, to replace a man in any kind of work[2]. Originally this robot was designed as a mechanism, which repeats motions after a human. For that a special motion capturing costume must be worn [3].

Authors have opted to use ROS and Gazebo software solutions.

ROS (Robot Operating System) is a widely used robot programming framework. It provides a mechanism for interprocess communications via named channels (topics) by typed messages [5]. It uses a "package" approach, when related software artifacts are packed together with their metadata. By using the standard message types AR-601 is able to utilize wide range of ROS packets, available "out of the box".
Gazebo is a physical and robotic simulator, excessively used in robotechnics. It provides user with a virtual world with simulated physics like gravity and solid bodies interactions [4].

AR-601 has its own proprietary UDP-based protocol, thus a protocol adapter was needed to translate between UDP and ROS messages.

This adapter for AR-601 to ROS and Gazebo was implemented by authors. Also, a Gazebo plugin was developed to mimic AR-601's peculiarities and characteristics. Both adapter and plugin implement the same basic interface, based on standard ROS messages with additional services for power control and motor mode changes. Additionally, some Python libraries were developed to ease further robot control programming. A strict robot definition in URDF format for ROS was created as well, to allow a proper integration with ROS.

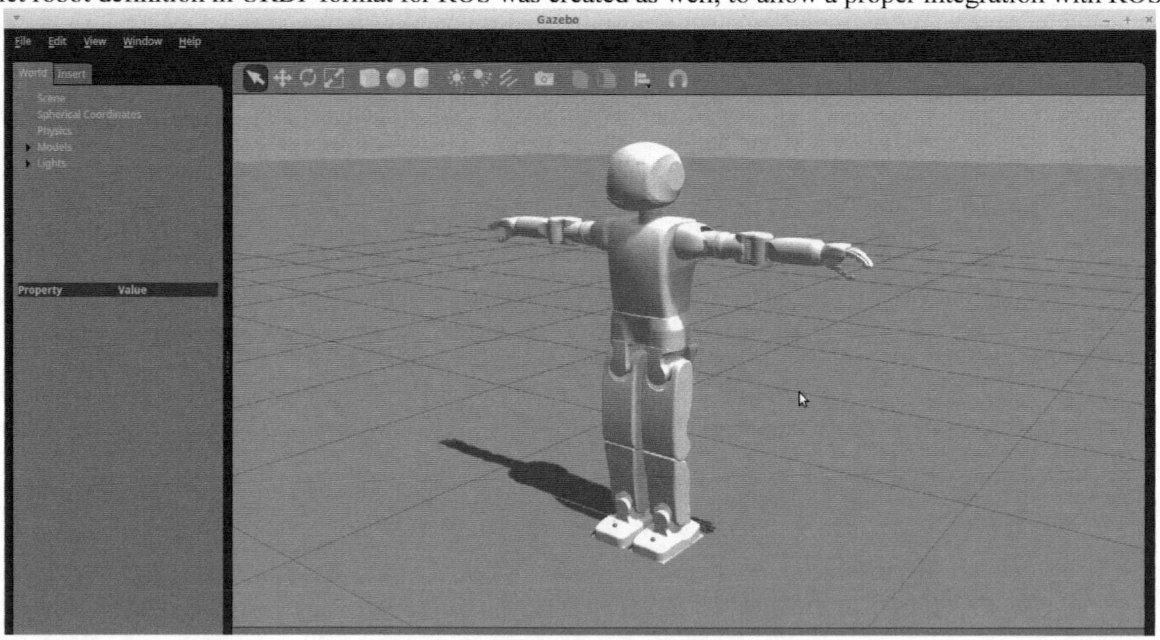

Fig. 2. URDF model in Gazebo.

Thus, rapid switch from model to actual robot is possible.
Retaining standard ROS message types enables a lot of other 3rd party ROS packages to interact with AR-601 and its Gazebo model. For example:
RViz for different visualization needs like sensor data, kinematic constrains, motor efforts and forces, etc
MoveIt for manipulator planning
The developed set of software artifacts, including urdf model, gazebo plugin and adapter might serve scientific as well as educational purposes for student's practical work in robotics. It allow to debug and test control programs on a virtual model and then quickly switch to real AR-601 without any code changes.

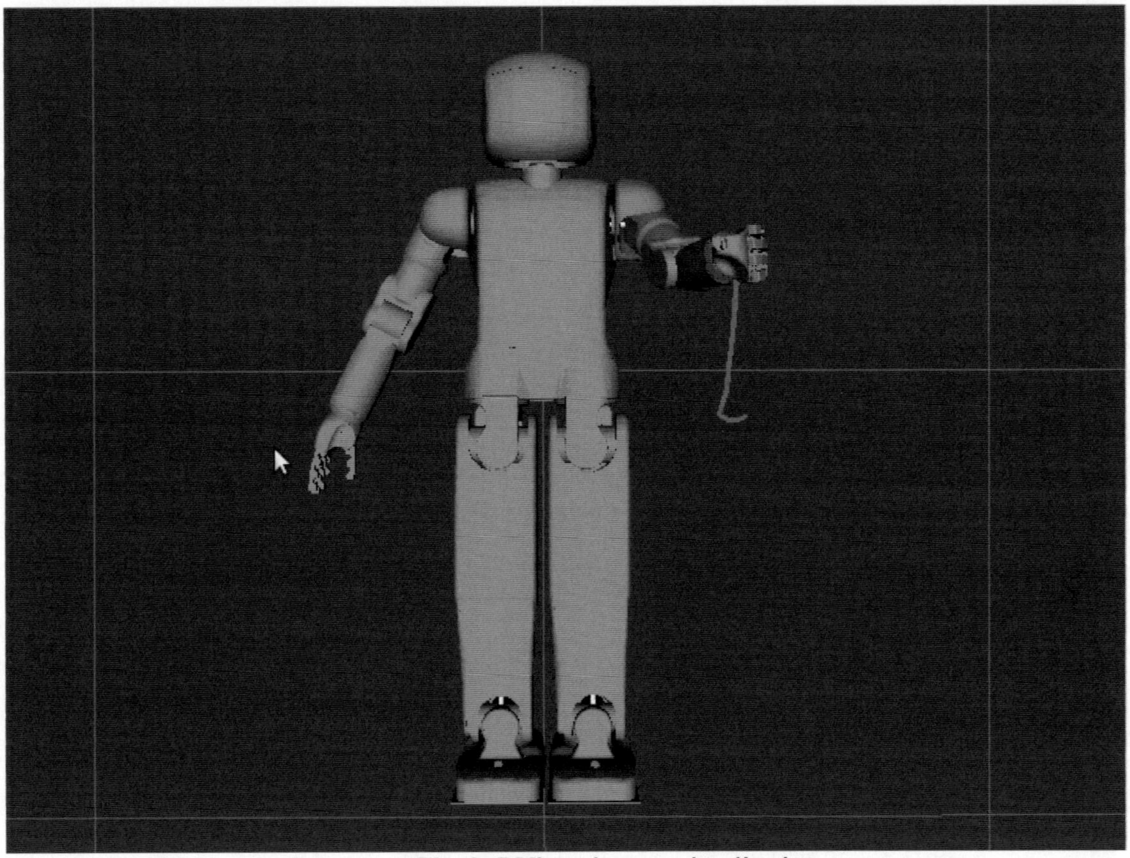

Fig. 2: RViz trajectory visualization

References:

[1] R Sirazetdinov, V Devaev, A Kamalov, E Katsevman 2015 *Program complex of modeling and virtualization anthropomorphic robot AR-601 bases on ROS and GAZEBO systems* IMMOD-2015, Moscow pp 328-331
[2] A Bogdanov, I Kutlubaev, A Permyakov, V Sichkov 2015 *Building an anthropomorphic robot including interaction control. Irreversible processes in nature and technology* Digest of Eighth-Russian Conference, pp 228-229
[3] NPO "Androidnaya technika":http://npo-at.com/products/ar-600e/
[4] Gazebo documentation: http://gazebosim.org
[5] ROS documentation: http://wiki.ros.org

IOP Conf. Series: Materials Science and Engineering **134** (2016) 012049 doi:10.1088/1757-899X/134/1/012049

Development control systems takeover: subject robotic arm on the example anthropomorhic robot AR-601

R Sirazetdinov[1], V Devaev[1], G Zakirzyanova[2]

[1] Kazan Federal University, Russia
[2] KNRTU-KAI them. Tupolev

rif-kat@inbox.ru, vdevaev@yandex, yaguzya97@gmail.com

Abstract. It is proposed the formation software architecture of complex motion for robotic systems in the form of sets of behavior - patterns - similar to the motor reflexes of living organisms. To form patterns of behavior of the robot teh authors used a methodology of structural analysis of complex systems IDEF0, then developed types of elementary algorithms (patterns) that make up the dynamics of the anthropomorphic robot jump. The work is performed according to the Russian Government Program of Competitive Growth of Kazan Federal University.

Management of live organisms based on the use of unconditioned and conditioned reflexes. Unconditioned reflexes are the body's reaction to the "typical" boring, to which the body responds to "standard" way. Unconditioned reflexes inherent in organism from birth and are constant for the entire lifetime. Conditioned reflexes are acquired as a result of training, are not innate, can be changed during the life [1].

An important role for the organism have the unconditional or somatic motor reflexes ensuring unconscious body's response to external disturbances is the basis for its motion, the protective reactions of defense. In this paper, we propose to use a reflex mechanisms for the development of control algorithms anthropomorphic robot. Use reflexes, similar to the motor reflex of humans and animals, will reduce the development time control algorithms for robot and make the robot more dynamic and resilient. Furthermore, it will implement parallelized solver algorithms to accelerate the formation of the robot control actions.

Name these reflexive mechanisms applied to traffic control robot behavior- patterns. For each of these patterns, you can define a set of measurable parameters of the robot current state, targets, characterized by a set of parameters of the state of governance and performance criteria.

The movement of the robot is regarded as a set of a large number of patterns, which are analogues of conditioned and unconditioned reflexes of a living organism. To determine the structure of this set will use the methodology of structural analysis SADT, have developed a certain methodology IDEF0 (Icam DEFinition). IDEF0 is a technology description of the whole system as a set of interrelated activities or functions [2].

In accordance with this methodology, any meaningful control of the robot can be considered as a series of nested SADT-diagrams, starting and finishing from the contextual diagrams describing the basic patterns of the robot behavior.

Form the patterns of the robot behavior used for walking on a flat surface, in accordance with the methodology IDEF0. Uses traditional circuit formation step anthropomorphic robot [3, 4].

1. At the top level description or consider contextual diagram step. The diagram in block form A0.

2. One of the basic functions is to manage the balance step to ensure the position of the robot body close to vertical. By analogy with human behavior control is carried out or stops, or hand movements or inclinations of the body. Since step precedes the state, then the diagram level 1 (Fig. 1) to ensure the balance indicated by block A1.

3. Step is carried out by transferring the non-reference legs to some preselected point on the surface - block A2 (Fig. 1).

4. After transfer of non-reference legs have to move it a foothold, as indicated on the diagram level 1 unit A3 (Fig. 1).

5. Next, the procedure is repeated non-reference transfer feet, which is not reflected in the charts.

6. The block for a balance A1 undergoes further decomposition is depicted in figure 2-level sequence of blocks A11, A12, A13 (Figure 2). It is at this level of decomposition pattern appears A13-stabilization.

Pattern A13-stabilization is a kind of unconditional reflex is not necessary to further decompose.

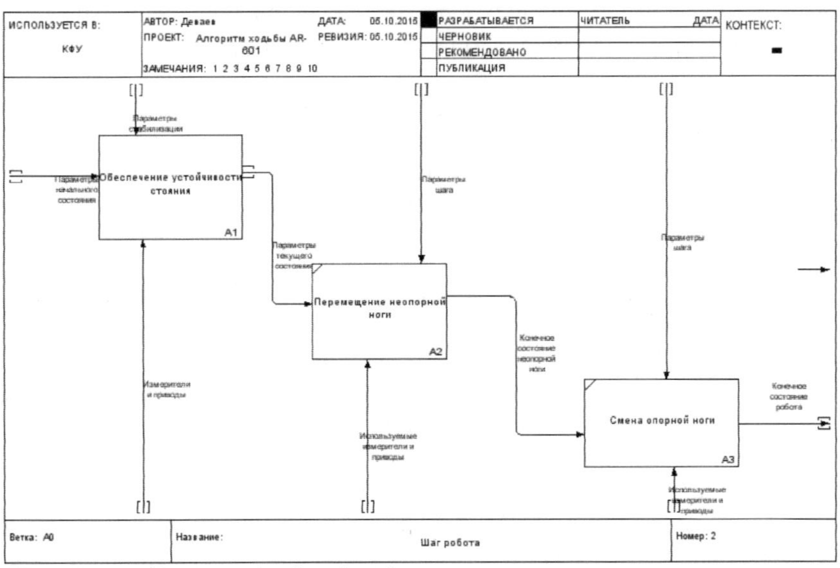

Fig.1. Steps diagram, the first level.

IOP Conf. Series: Materials Science and Engineering **134** (2016) 012049 doi:10.1088/1757-899X/134/1/012049

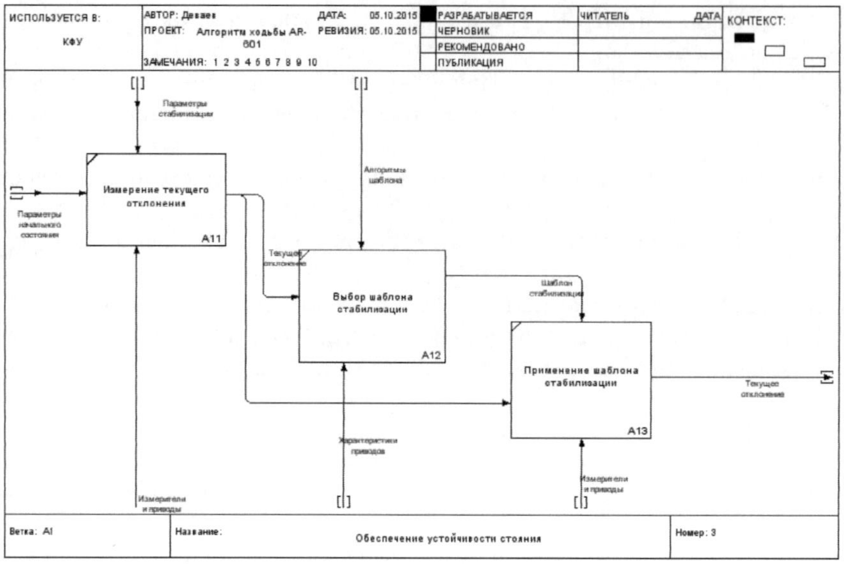

Fig. 2. Steps diagram, a second level.

Thus, this paper proposes and illustrates the application of the structural analysis method of complex processes control movement of robot based on the methodology IDEF-0 to highlight the elementary behavior patterns that are later implemented in software.

An example of the stabilization implementation of the template is given in [5], which shows the stabilization of the anthropomorphic robot AR-601 in the longitudinal plane mode state by controlling the drive stops and the measurement deviation from the vertical by a strain gauge force-torque sensors foot.(See Fig. 3) Also, an algorithm is implemented hopping robot in the following patterns.

1. The pattern squats. It is performed by the instructions from outside the robot. The depth squats determined allowable acceleration, which can then straightens his legs.

2. The pattern of leg extensions. It is activated once the pattern completes squats. Straightening the legs is performed so that the gap of the robot from the floor surface took place at a predetermined speed. The magnitude and direction of the speed determined in the future trajectory of the robot jump.

3. The pattern of balancing during the flight. Performs small hand movements, stabilizing the movement along the calculated trajectory.

4. The pattern of the landing (squat). It is activated when the robot touches the floor. Measured overload and done squat to ensure permissible overload. The pattern in passive mode when the magnitude of the overload becomes close to the acceleration of free fall.

5. Pattern straightening the legs. Activated at the close of the previous pattern.

6. The pattern of static stability. It is active when is a load of the feet of the robot. During the flight, it is deactivated. This pattern provides stability by turning the robot drives the ankle and is designed to prevent the fall during the remaining patterns.

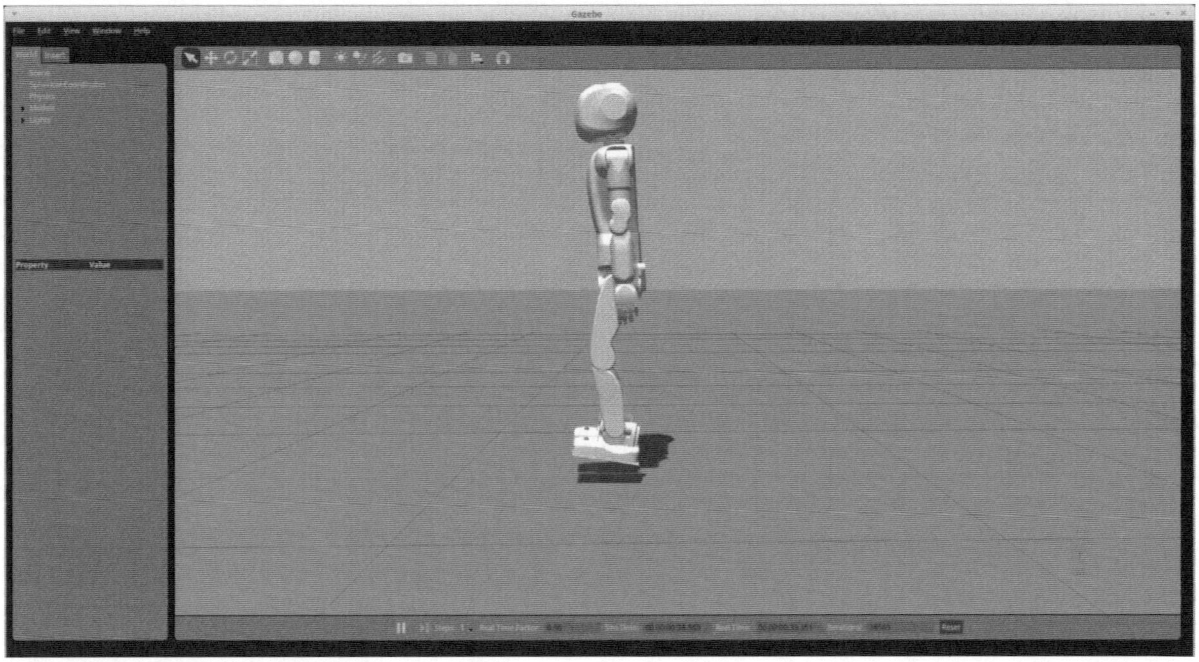

Fig. 3. Robot jump pattern.

References.

[1] H Danilova, A Krylov 2005 Physiology of Higher Nervous Activity - Rostov ,: "Phoenix" S Cheremnyh, I Semenov, B Ruchkin 2006 - M Modeling and analysis systems. IDEF-Technology: Workshop : Finance and Statistics
[2] V Beletsky1984 Bipedal walking: modeling problems of dynamics and control
[3] P Lozhkyn, O Tolstel 2013 Create hardware and software, providing a dynamic balance and movement of RTC AR-600 Herald Baltic Federal University Vol 4 pp 110-117
[4] D Nikitina, A Fadeev 2015 Development of algorithms for dynamic walking on an example of the anthropomorphic robot AR601. Eighth All-Russian Conference of Young Scientists